Environmental Rhetoric and Ecologies of Place

Environmental Rhetoric and Ecologies of Place explores some of the complex relationships, collaborations, compromises, and contradictions between human endeavor and situated discourses, identities and landscapes, social justice and natural resources, movement and geographies. This collection serves to "unpack" and grapple with the complexities of rhetoric of presence and addresses the complexities of rhetorical praxis in relation to place. Through case analyses, these essays examine and illustrate the concepts and practices of knowledge making and knowledge distribution at confluences of geographical and geospatial locations as well as human perceptions, experiences, and interpretations of the world, nature, and the environment. These essays challenge us to ponder our futures in the theoretical and lived places we inhabit. In focusing on regional environmental issues, this collection offers a corrective to what appears an increasingly hegemonic discourse of globalization that conceives of the world as flattened.

Peter N. Goggin is Associate Professor of English (Rhetoric) at Arizona State University, USA

Routledge Studies in Rhetoric and Communication

1 Rhetorics, Literacies, and
Narratives of Sustainability
Edited by Peter Goggin

2 Queer Temporalities in Gay
Male Representation
Tragedy, Normativity, and Futurity
Dustin Bradley Goltz

3 The Rhetoric of Intellectual
Property
Copyright Law and the Regulation
of Digital Culture
Jessica Reyman

4 Media Representations of
Gender and Torture Post-9/11
Marita Gronnvoll

5 Rhetoric, Remembrance, and
Visual Form
Sighting Memory
*Edited by Anne Teresa Demo
and Bradford Vivian*

6 Reading, Writing, and the
Rhetorics of Whiteness
Wendy Ryden and Ian Marshall

7 Radical Pedagogies of Socrates
and Freire
Ancient Rhetoric/Radical Praxis
S.G. Brown

8 Ecology, Writing Theory, and
New Media
Writing Ecology
Edited by Sidney I. Dobrin

9 The Rhetoric of Food
Discourse, Materiality, and Power
*Edited by Joshua J. Frye
and Michael S. Bruner*

10 The Multimediated Rhetoric of
the Internet
Digital Fusion
Carolyn Handa

11 Communicating Marginalized
Masculinities
Identity Politics in TV, Film,
and New Media
*Edited by Ronald L. Jackson II
and Jamie E. Moshin*

12 Perspectives on Human-Animal
Communication
Internatural Communication
Edited by Emily Plec

13 Rhetoric and Discourse in
Supreme Court Oral Arguments
Sensemaking in Judicial Decisions
Ryan A. Malphurs

14 Rhetoric, History, and Women's
Oratorical Education
American Women Learn to Speak
*Edited by David Gold
and Catherine L. Hobbs*

15 Cultivating Cosmopolitanism for
Intercultural Communication
Communicating as Global Citizens
*Miriam Sobré-Denton
and Nilanjana Bardhan*

**16 Environmental Rhetoric and
Ecologies of Place**
Edited by Peter N. Goggin

Environmental Rhetoric and Ecologies of Place

Edited by Peter N. Goggin

Routledge
Taylor & Francis Group
NEW YORK   LONDON

First published 2013
by Routledge
711 Third Avenue, New York, NY 10017

Simultaneously published in the UK
by Routledge
2 Park Square, Milton Park, Abingdon, Oxon OX14 4RN

*Routledge is an imprint of the Taylor & Francis Group,
an informa business*

© 2013 Taylor & Francis

Library of Congress Cataloging-in-Publication Data
  Environmental rhetoric and ecologies of place / edited by Peter N. Goggin.
    pages cm. — (Routledge stuides in rhetoric and communication ; 16)
    Includes bibliographical references and index.
  1. Communication in ecology. 2. Communication in the environmental sciences. I. Goggin, Peter N.
  QH541.18.E68 2013
  577—dc23
  2013007158

ISBN13: 978-0-415-81840-7 (hbk)
ISBN13: 978-0-203-54941-4 (ebk)

Typeset in Sabon
by IBT Global.

SUSTAINABLE FORESTRY INITIATIVE

Certified Sourcing
www.sfiprogram.org
SFI-01234
SFI label applies to the text stock

Printed and bound in the United States of America
by IBT Global.

**For the islands
and all that dwell there**

Contents

List of Figures   xiii
Acknowledgments   xv

Introduction   1
PETER N. GOGGIN

PART I
Places We Dig (Mine)

1   A Certain Uncertainty: Drilling Into the Rhetoric of the
     Marcellus Shale Natural Gas Development   15
     JAMES GUIGNARD

2   Eco-Seeing a Tradition of Colonization: Revealing Shadow
     Realities of Marcellus Drilling   28
     BRIAN COPE

3   Sense of Place, Identity, and Cultural Continuity in an Arizona
     Community   42
     DEBORAH L. WILLIAMS AND ELIZABETH A. BRANDT

4   Mt. Taylor, New Mexico: Efforts to Provide Resilience to a
     Sacred Mountain Socio-Ecological System   54
     SALLY SAID

PART II
Places We Build and Create

5   A Land Ethic for Urban Dwellers   69
     GESA E. KIRSCH

6 "We Face East": *The Narragansett Dawn* and Ecocentric
   Discourses of Identity and Justice   84
   MATTHEW ORTOLEVA

7 Conjuring the Farm: Constructing Agricultural Places in U.S.
   Schools   97
   CYNTHIA R. HALLER

8 Digital Cities: Rhetorics of Place in Environmental Video
   Games   111
   MICHAEL SPRINGER AND PETER N. GOGGIN

PART III
Places We Travel Through, Around, and Within

9 Reading the *Atlas of the Patagonian Sea*: Toward a Visual-
   Material Rhetorics of Environmental Advocacy   127
   AMY D. PROPEN

10 A Place of One's Own   143
   SAMANTHA SENDA-COOK AND DANIELLE ENDRES

11 Local Flaneury: Losing and Finding One's Place   155
   JAQUELINE MCLEOD ROGERS

PART IV
Places of Resistance and Acceptance

12 From Concept to Action: Do Environmental Regulations
   Promote Sustainability?   173
   BECCA CAMMACK, LINN K. BEKINS, AND ALLISON KRUG

13 Mapping Literacies: Land-Use Planning and the Sponsorship
   of Place   186
   REBECCA POWELL

14 Place-Identity and the Socio-Spatial Environment   199
   RICK CARPENTER

Afterword   217
KIM DONEHOWER

Contributors   225
Index   231

Figures

5.1   View of the Longfellow Bridge from the Prudential Tower observatory.   72

5.2   The Piers (Salt and Pepper Shakers). Longfellow Bridge Between Cambridge and Boston, Massachusetts, USA (Daderot).   73

7.1   "Keys to Stewardship" reward map.   102

9.1   Regional nesting sites of the Wandering Albatross (Diomedea exulans).   131

9.2   Wandering Albatross (Diomedea exulans): principal feeding areas in the Patagonian Sea.   133

9.3   Elephant Seal with satellite tracking device.   135

14.1   Protestors in downtown Valdosta.   206

Acknowledgments

The idea for this collection emerged from a confluence of events, experiences, and influences that served to highlight the need for scholarship that illustrates the role of rhetorical analysis in exploring how we understand environmental concerns in particular locales—especially in a seeming climate of globalization discourses that presume to eradicate notions of place. I would first like to acknowledge participants for the 2011 Western States Rhetoric and Literacy conference, *Places*, who demonstrated over two days of sharing conversations and presentations just how important the concept of place is for scholarly inquiry. In addition, I would like to acknowledge my graduate students in my courses on Environmental Rhetoric and Sustainability and my undergraduates in my courses on English Studies and the Environment whose enthusiasm for learning keeps the excitement for inquiry in this area vibrant and purposeful. And along with them, my thanks to my colleague Ron Dorn for his collaboration on field trips and teaching about environmental sustainability. To my colleagues in the Global Institute of Sustainability, and Ann Kinzig in particular, I want to thank them for our think-tank sessions on wicked problems and questions concerning the future for local and global concerns. My thanks also to Routledge/Taylor Francis editors Liz Levine, Emily Ross, and editorial assistant Nancy Chen for their support and guidance on this project, to the reviewers for their critical and thoughtful input, and to the all the contributors whose cooperation, participation, and efficiency in putting this collection together made the project a pleasure to work on. Most importantly, my thanks to Maureen Daly Goggin for her ongoing support for creative research and teaching as Chair of ASU's Department of English, and as my spouse for her generous support and feedback throughout the process of compiling and editing this collection. Finally, my thanks to all the people in the many places I've been privileged to visit, experience, teach in, and study who underscore the significance of places, their environments, and our relationships to them.

Introduction

Peter N. Goggin

A place has a human history and a geologic past: it is a part of an eco-system with a variety of microsystems, it is a landscape with a particular flora and fauna. Its inhabitants are part of a social, economic, and political order: they import or export energy, materials, water, and wastes, they are linked by innumerable bonds to other places. A place cannot be understood from the vantage point of a single discipline or specialization. It can be understood only on its terms as a complex mosaic of phenomena and problems. (Orr 129)

Instead then, of thinking of places as areas with boundaries around, they can be imagined as articulated moments in networks of social relations and understandings, but where a larger proportion of those relations, experiences and understandings are constructed on a far larger scale than what we happen to define for that moment as the place itself, whether that be a street, or a region or even a continent. (Massey 154)

Give me a place to stand, and a lever long enough, and I will move the Earth. (Archimedes)

In a frequently cited passage in *Everybody's Autobiography*, Gertrude Stein writes about visiting her childhood home: "Anyway what was the use of my having come from Oakland; it was not natural to have come from there. Yes write about it if I like or anything if I like but not there, there is no there there" (289).

On August 6, 2012, the National Aeronautics and Space Administration (NASA) landed the robotic rover science lab, Curiosity, on the surface of the planet Mars. The mission's goal: to seek out evidence of organic carbon to determine if the planet may have ever supported microbial life.

As with the epigraphs above, these two cases represent just how broad the spectrum is on "place" as a rhetorical construction. In Stein's case, place is a representation of individual identity and remembered experiences of material things and locations. The space that is filled with her memories of the material reality of what she knew as Oakland remains, but the material that constitutes a sense of place for her no longer exists. Of course, for the population of Oakland, California, there is a "there" there, and the paradox of Stein's observation on place is to give the place even more thereness through her rhetorical construction of Oakland in print and the public consciousness.

On another side of the spectrum, the Mars rover project illustrates an evolving cultural shift from the notion of place as intangible speculation to one of material actuality. The Mars rover/lab, Curiosity, is on the surface of that planet and is gathering physical and instrumental data. If there is evidence that life ever existed on Mars, the study of this planet's past changes in climate may also contribute to comparative analysis for understanding climate change on Earth. Further, the data the rover collects will contribute to the potential for humans to live on Mars and ultimately exploit its resources. This planet—that for most of human awareness has been a place of myth, religious mystery, and fiction—has become a material reality. As Kim Stanley Robinson states about colonization of this planet in his futuristic novel *Red Mars*, "And so we came here. It had been a power; now it became a place" (3).

Although these two variations of perspective would appear to discount each other, place as geographically ephemeral constructed memory and place as geocentric material presence, they complement each other in terms of seeing place as something transformative. In *The Fate of Place*, Edward Casey's rich historical synthesis of place and space in Western philosophy, the author observes that from about the seventh century the concept of place has been subsumed by obsession with space to the point of relegating place to obscurity in philosophical thought by the eighteenth century. Casey proposes that the spread of Christianity and the notion of infinite space, the universal logic of pure mathematics, universal linguistics, and the impact of colonialism and world wars in eroding any durable security in people's sense of place are likely causes of this phenomenon. However, he argues that the contemporary encroachment of globalization as a pervasive concept and presumption of inevitable sameness has renewed place as a viable subject of reflective thought. Stein and Curiosity, and the epigraphs at the beginning of this introduction, remind us that concepts of place are emerging as a crucial context for humanistic inquiry because of a growing acceptance that human activity has lasting impacts on the earth's ecosystems (in geological terms we are now in an era some scientists are calling the Anthropocene Epoch). It is a recognition that the quality of life for future generations is at stake depending on what we, as societies and civilizations, choose to enact now in terms of economic, environmental, and social development, in global and local spheres.

The collection of essays in *Environmental Rhetoric and Ecologies of Place* addresses the complexities of rhetorical praxis in relation to place. Through case analyses, these essays examine and illustrate the concepts and practices of knowledge making and knowledge distribution at confluences of geographical and geospatial locations as well as human perceptions, experiences, and interpretations of the world, nature, and the environment. These essays challenge us to ponder our futures in the theoretical and lived places we[1] inhabit. In focusing on regional environmental issues, this collection offers a corrective to what appears an increasingly hegemonic discourse of globalization that conceives of the world as flattened. Geopolitics, global economics, and communication and transportation technologies are but a few of the dominant forces in our world that presume to inform a perspective in which place itself is no longer relevant as political, socio-cultural, and economic systems are integrated. It is an ideal world where sustainability of natural and synthetic ecologies can be realized through globally ordered socio-economic utopia and/or Promethean epistemology based on the promise of technological innovations and solutions.

Place is a concept of human value where globalization is seen and welcomed for many as a signifier of the redundancy of place as geospatial identity becomes increasingly irrelevant to communication and commerce, and as transnational corporations replace nation-states as the source of power, security, and leadership in the world, what ecological literacy theorist David Orr refers to as the end of the Westphalian war system. In contrast to natural systems, he states, human society "is increasingly homogeneous. The great diversity of human cultures is being rapidly destroyed by the global force of modernization. Humans are now being congealed into one great megaexperiment" (59). Other scholars make the case that

> globalization—especially as driven by the revolution in information and communications technologies (ICT)—marks the "end of geography" (O'Brien, 1992), the onset of the "death of distance" (Cairncross, 1997), the emergence of a 'borderless world' (Ohmae, 1995), of "de-territorialization" or "supra-territorialization" (Scholte, 2000) and the "vanishing of distance" (Reich, 2001) . . . [and] that as a consequence of globalization, "the world is flat" (Friedman, 2006) (Christopherson, Harry Garretsen, and Ron Martin 343)

For economic purposes of global trade and infrastructure management that foster and maintain the production, distribution, and consumption of goods and services, the eradication of place other than as constructed simulacra makes good sense. We expect our chain stores, automobile dealerships, fast food restaurants, and so forth to be essentially identical nationally and worldwide. Browsing a shopping mall in Des Moines is pretty much like browsing in a shopping mall in Los Angeles or Dubai or London. The daily process of commuting to work, grocery shopping, watching *American Idol* (or the national franchise version) in the evening or playing *Angry Birds* is remarkably universal for many people living in (most) developed countries.[2] Regarding environmental concerns, climate change, species extinction, pollution, water, and food production are just a few issues the population of the whole planet shares, and require global long term solutions.[3] Global Complexity theorist John Urry states that "the analysis of globalization brings out the obvious interdependencies between peoples, places, organizations and technological systems across the world." However, Urry concludes this observation by observing "with the analysis of globalization, 'no place is an island'" (39).

The general intent of Urry's statement on global complexity is well taken. There is likely no place on earth where there is human settlement where daily life and cultural values are not impacted on in some way by global connection. But it seems that this reality is all too often used to rationalize the unfettered exploitation of resources and people in the name of "good" for society, often while invoking a conveniently worded mantra of sustainability (see Goggin). The environmental impacts of global commerce are always located some place that is real and discrete. Within complex global systems the "no places" Urry refers to are actual islands, neighborhoods, apartment blocks, villages, farms, forests, savannas, and so forth, and although, yes, these locales are enmeshed in the global network of commerce, politics, communication, and ecologies, they are, in fact, places in their own right. Here, Urry illustrates an important limitation of global perspective in both popular and scholarly discourse that situates discrete places and place perspectives as outside of a dominant, homogenous global reality.

In their analysis of rural rhetorics and literacies, critics of global homogeneity Donehower, Hogg, and Schell point to a phenomenon of commonplace assumptions constructed through an urban/mainstream worldview they term "rurality." Sheller and Thompson equate a similar phenomenon for oceanic islands (the Caribbean in particular) that is in large part informed by colonial and postcolonial images that have defined island natural landscapes, the people, and their cultures as picturesque, and thus relegated as consumable places within the global system. Drawing on Pratt's study of imperial travel, Sheller observes that such a view constitutes a "rhetoric of presence" that fixes "the mastery of the seer over the seen" (50). In these cases, the perception of rural and island people, places, and ecosystems constitutes a form of "world making" that reinforces a sense of timeless dissonance for the mainland/mainstream worldview. In her 2012 edited special issue of *Rhetoric Society Quarterly* on "Regional Rhetorics," Jenny Rice observes that "what is particularly insidious about flat data is that it smoothes over the tectonics of place. The light, the air, the stubbornness of any given landscape is eradicated by a construction that makes everything level" (202). For these scholars, as proponents of the significance of discrete places and locales, accepted and unreflexive constructs of globalization perpetuate a perspective that diminishes the existence of places that otherwise offer alternative or, at least, subtlety nuanced viewpoints on the world.

ENVIRONMENTAL RHETORIC

Understanding how rhetoric, and environmental rhetoric in particular, informs and is informed by local and global ecologies contributes to our conversations about sustainability and resilience—the preservation and conservation of the earth and the future of human society. *Environmental*

Rhetoric and Ecologies of Place is designed to explore some of the complex relationships, collaborations, compromises, and contradictions among human endeavors and situated discourses, identities and landscapes, social justice and natural resources, and movement and geographies. In short, this collection serves to "unpack" and grapple with the complexities of rhetoric of presence and place.

As a subject of scholarly inquiry in the humanities, and in rhetoric in particular, a number of scholars have made significant contributions in their work on environmental rhetoric. For example Killingsworth and Palmer in *Ecospeak*; Herndl and Brown (eds.) in *Green Culture*; Dobrin and Weisser in *Natural Discourse*; Muir and Veenendall in *Earthtalk*; Owens in *Composition and Sustainability*; Tarla Peterson in *Sharing the Earth*; and Donehower, Hogg, and Schell in *Reclaiming the Rural*, to name a few. An emphasis on environmental and ecological concerns enjoys growing interest in the public sphere and growing scholarly legitimization as an emerging area of inquiry across multiple academic fields and disciplines. Along with this growing interest is recognition of the complexities of competing ideologies that inform the discourses of globalization, localization, and glocalization as well as the confluences, collisions, and fusing that occur in the liminal spaces and places where technological and ecological systems intersect and interact, whether through natural or human-caused events. Increasingly we are faced with the dynamics of scaling in our predictions and responses to issues of place as they impact on us and our world on local, regional, and global strata.

But although the purpose of this collection is to underscore the importance of place-based rhetorics on the environment and make a case for the ways that rhetorical analysis of place can contribute to visions and arguments concerning a resilient and sustainable future, one must be careful not to romance the local. Environmental rhetorics, such as those rooted in epistemologies of deep ecology, and other forms of ecocentric environmental activism tend to contribute to polarizing public opinion on environmental concerns and engender divisive discourse that Killingsworth and Palmer term "ecospeak." As problematic as globalization, which presumes to eradicate the nuances of place, localization presumes a relativistic worldview that emanates from an individual or location based core, a sort of cultural universalism. James Cantrill, for example, in an analysis of environmental investment and sense of place finds that the longer individuals have dwelt in a particular place or region, the more they tend to value social relations over natural environment, particularly if there is a perception that promotion of ecosystem management may impact on immediate self-interests. If globalization presents the world as flat, then localization presents a view that funnels global terrain through a parochial lens like Saul Steinberg's famous, and much copied 1976 cover of *The New Yorker* magazine, "View of the World from 9th Avenue." Rhetorics that reframe (some may say, reclaim) a worldview that sees place in terms of *glocalization* challenge the binary perspective of the global versus the local, and offers insights into how rhetorics of place can be employed in a more robust understanding of the interconnected complexities of both global and local participation in ecological sustainability. Deliberative rhetoric, for example, does not presume to seek consensus per the classical understanding of deliberation, but rather offers such a reframing of place and environmental discourse by encouraging stakeholders to articulate opinions and beliefs as a means to critically examine these (and those of others) to foster effective argumentation, change, and synchronicity (Coppola and Karis; Scialdone-Kimberley and Metzger; Said). Likewise, regional rhetorics offer "a response to the immediate tectonics of history, embodiment, and physical shapings of materiality . . . disrupts given narratives of belonging that are framed on a national level and between individuals. Regional rhetorics provide alternative ways of framing our relationships and modes of belonging" (Rice 203). Such rhetorics as these (deliberative and regional) offer zones and seams of connection (what anthropologist Anna Lowenhaupt-Tsing refers to as "frictions") between local universal knowledges and global hegemonic perspectives, offering an opportunity to unflatten the terrain and provide a nexus for discourse that resists geophysical relativism.

ECOLOGIES OF PLACE

Generally defined, ecology is the interdisciplinary scientific study of relationships among communities of plants and animals (organisms) and their environments. This definition encompasses a vast array of interactive systems, from microscopic ecosystems to entire planetary systems. Although ecology is not necessarily concerned with the preservation of environmental systems and tends to focus on the interactivity along with evolved and evolving of organisms in relation to their environments, the findings of this science are crucial to the preservation and conservation of biodiversity. However, the generalizing and applicability of deep knowledge of a particular ecological system to the broader issues of environmentalism is problematic. As biologists Ian Billick and Mary Price note in their introduction to their collection *The Ecology of Place*, "diversity and contingency present enormous challenges to understanding the ecological systems of the planet because there are simply too few ecologists to study them all" (1). The editors go on to argue that one "obvious approach" to addressing the problem of generalizing ecological analysis is through comparison "across taxa[4], space, and time." They go on to name three viable options, "the development and testing of general-theory—broadly applicable testing propositions . . . the quantitative description of predictive empirical patterns uncovered by the comparative approach, or the elaboration of logical consequences of empirically or deductively generated premises;" and generalizations that "involve a statement about the applicability of alternative conceptual frameworks to a particular system with particular properties" (3–4). This third option they define as "ecology of place," Billick and Price draw on "because that approach pursues general understanding through the sort of detailed understanding of a particular place—the 'sense of place'—that has come to be associated with ecologist and conservationist, Aldo Leopold" (4). They state,

> We reserve "place" to represent all of those idiosyncratic ecological features—including spatial location and time period—that define the ecological context of field study; and "ecology of place" or "place-based research" for research that assigns the idiosyncrasies of place, time, and taxon a central role in its design and interpretation, rather than as a problem to be circumvented through replication or statistical control. (5)

Geologist Steven Semken further defines notions of place and space, and the concept of sense of place as imbued with social science and humanist ideals that denotes "the meanings and the attachments to a place held by a person or a group" (149). He states the following:

> Place is distinguished from space by being socially constructed and local rather than quantitatively described and universal (Tuan, 1977). In other words, people make places out of space (Brandenburg and Carroll, 1995), and a given locality or landscape can hold widely divergent meanings for different individuals or cultures (Gruenwald, 2003). The physical environment appears to play a major role in creating and shaping sense of place (Ryden, 1993; Stedman, 2003), and in some cases a physiographic province or ecosystem may coincide with a place (Williams and Patterson, 1996). (149)

This collection takes a rhetorical approach to ecology as a metaphor and organizing principle for examining relationships between people and the natural, synthetic, and social systems of the places they dwell in. It is metaphorical in the sense that this is not a scientific approach that is commonly associated with fields of ecological research with their various methods of biotic systems analysis. Rather, the rhetorical approach to place in this collection draws from the concept of ecological inquiry as defined by Billick and Price, and the interpretation of environment by people in local contexts as described by Semken that examines how close analysis and understanding of local systems can be applied more broadly to generalizations of larger systems, and *vice versa*.

Environmental Rhetoric and Ecologies of Place brings together essays that explore the complex discursive constructions of environmental rhetorics and place-based rhetorics including the following:

- Discourses, actions, and adaptations concerning environmental regulations and development, sustainability, exploitation, and conservation of energy resources
- Arguments on cultural values, social justice, environmental advocacy, and identity as political constructions of rhetorical place and space
- Ethics and identities of environment in rural and urban case studies of place; rhetorics of environmental cartography and glocalization
- Rhetorical framing of the environment in terms of perspective and sense of place

This collection takes us beyond the assumption that rhetorics are situated, and challenges us to consider not only how and why they are situated, but also what we mean when we theorize notions of situated, place-based rhetorics.

The contributors to this collection represent a range of specialization across a variety of scholarly research in such fields as communication studies, rhetorical theory, social/cultural geography, technical/professional communication, cartography, anthropology, linguistics, comparative literature/ecocriticism, literacy studies, digital rhetoric/media studies, and discourse analysis. They offer diverse perspectives and methodologies that help to tease out the nuances in the issues of place and ecology. *Environmental Rhetoric and Ecologies of Place* is organized into four parts that are associated with common human interactivity and identity in the places we dwell across overlapping local, national and global contexts: resources, infrastructures, movement, and tension. These organizational categories are themselves rhetorical, and are not presumed to be exclusive. In fact, some of the essays located under one heading easily lap over into others and all could potentially be rearranged to create new categories (or none for that matter). Nevertheless these essays come together in an organizational scheme based on ways that they resonate with, or challenge, each other in their theoretical frames, their dominant themes, and critical perspectives. The essays in the first part, "Places We Dig (Mine)," address rhetorical conundrums—environmental, social, and spiritual—associated with the human need for natural mineral resources that must be extracted from the very ground we live on. The first two chapters by James Guignard ("Drilling Into the Rhetoric of the Marcellus Shale Natural Gas Development") and Brian Cope ("Eco-Seeing a Tradition of Colonization: Revealing Shadow Realities of Marcellus Drilling") deal with "fracking," the highly controversial practice of hydrofracturing deep underground shale beds to extract natural gasses. As a practice that has been banned in some countries and has been associated with ground water contamination, air pollution, and even earth tremors, fracking has been both welcomed by local communities and state governments in the United States as an economic windfall for depressed economies, and vilified by environmentalists and local residents concerned for the long-term cultural and physical impacts on the places they call home. The essays by Guignard and Cope illustrate the complexities of place-based stakeholder discourses in Pennsylvania locales on this highly polarizing issue. Likewise, the essays by Deborah Williams and Elizabeth Brandt ("Sense of Place, Identity, and Cultural Continuity in an Arizona Community") and Sally Said ("Mt. Taylor, New Mexico: Efforts to Provide Resilience to a Sacred Mountain Socio-Ecological System") take us to the town of Superior, Arizona and to Mount Taylor near Grants, New Mexico to explore copper mining and uranium mining, respectively, and the rhetorical conflicts that arise in small multicultural and indigenous communities in the debates over economic, cultural, and environmental sustainability in these places.

From ecologies of resources to ecologies of construction, the second part, "Places We Build and Create," explores cases that illustrate the discussions and conversations associated with the physical, textual, and digital infrastructures that mark and define notions of place. Gesa Kirsch in "A Land Ethic for Urban Dwellers" re-inscribes Aldo Leopold's concept in contemporary deliberations of place-based visions for sustainable urban futures in Boston, Massachusetts and the renovations of the Longfellow Bridge. In, "'We Face East': *The Narragansett Dawn* and Ecocentric Discourses of Identity and Justice," Matthew Ortoleva examines the impact of spiritual/environmental discourse in the 1930s publication of a Native American tribal newsletter in influencing an emerging counterpublic and creating a sense of ecological identity in the Narragansett Bay watershed in the southern Massachusetts/eastern Rhode Island region. Cynthia Haller, in "Conjuring the Farm: Constructing Agricultural Places in U.S. Schools," and coauthors Michael Springer and Peter Goggin, in "Digital Cities: Rhetoric of Place in Environmental Video Games," examine a growing area of pedagogical interest (and influence) associated with online and Internet-based corporate sponsorship equated with place-based learning—the creation of digital "places" that purport to educate on matters of environmental sustainability. In Haller's case, she examines rhetorical representations of ecologically positive farming published on a website published by the American Farm Bureau Foundation for Agriculture, whereas Springer and Goggin offer a case analysis of the ideological and rhetorical frames of the digital environments of *Energy City* (The Jason Project) and *Energyville* (Chevron).

Places are not static; and neither are the organisms and environments that they are made of. The third part, "Places We Travel Through, Around, and Within," gives a nod to the universal rule of entropy and addresses the kinetic aspect of movement in various ecological systems and environments. In Amy Propen's "Reading the *Atlas of the Patagonian Sea*: Toward a Visual-Material Rhetorics of Environmental Advocacy," the author examines visual and material rhetorics in the construction of the *Atlas* through data gathered over a decade of tracking the movements of marine species in the South Atlantic. Through these rhetorics, Propen argues, the *Atlas* makes a cartographic case for species agency and conservation. In "A Place of One's Own," Samantha Senda-Cook and Danielle Endres use rhetorical criticism to analyze outdoor recreation texts (magazines and catalogues) and interviews with hikers to explore and understand the implications of seeking solitude in nature—a practice, they argue, that preserves the rhetorical and physical nature/culture divide and, paradoxically, may cause degradation to the very places nature recreators seek out. In her essay, "Local Flaneury: Losing and Finding One's Place," Jaqueline McLeod Rogers argues for the practice of alert, socially and culturally aware urban walking based on a reconstruction of the French notion of the *flâneur*, an aesthetically attuned yet socially detached observation of urban landscapes, for teaching a reinventing of city space and place.

Although all of the chapters deal with challenges, complexities, and controversies in rhetorics of places, the fourth section, "Places of Resistance and Acceptance," includes essays that examine the theoretical and deliberative considerations of place-based rhetorics that inform activism and citizenship in the interests of environmental and social sustainability. In "From Concept to Action: Do Environmental Regulations Promote Sustainability?" Becca Cammack, Alison Krug, and Linn Bekins conduct an analysis of interviews with San Diego County water utility employees to examine how national and state environmental regulations impact corporate and personal/local choices about sustainability, and if place itself is a contributing factor in effecting action. Their study offers insights on the rhetorical power of regulatory discourse to influence a citizenry that accepts a view of environmental sustainability as concept to shift to an acceptance of sustainability as culture. Rebecca Powell, in "Mapping Literacies: Land-Use Planning and the Sponsorship of Place," examines community deliberations on a vision for regional planning in Doña Ana County, New Mexico and argues for the importance of community literacy and civic participation based on the lived experiences of county residents to foster a sense of place in environmental activism. The final chapter in this collection, "Place-Identity and the Socio-Spatial Environment" by Rick Carpenter, is a case analysis of geographic rhetoric employed by residents of Valdosta, Georgia to employ place-based arguments in opposition to a proposed biomass incinerator plant. In her afterword, Kim Donehower draws on Scottish philosopher Alasdair MacIntyre's treatment of moral rationality and discourse as a provocative lens for framing the essays on place in this collection as indicative of the competing multiple and complex perspectives that clash and compete for dominance in arguments on environmental concerns. Together, this collection of essays provides scholars a window into the complex and often contradictory arena of discourse on the environment and illustrates some of the contemporary theory and research on place-based rhetorics in this important, emerging area of inquiry in the humanities.

NOTES

1. In this introduction and the table of contents I intend "we" (and "our") to be provocative in evoking questions about who is present and not present,

included and not included, enfranchised and disenfranchised in connections through place, but I also evoke the ecological "species" view of humanity and all other life in the biotic community and global systems of Earth's biosphere.
2. "Angry Birds was recognized as the top paid-for App Store game in most countries, having sold more than 6.5 million downloads of the iOS version of the game since it was first released in December 2009" ("Apple App Store").
3. For example, on the issue of food production, the *Sustainable P Initiative* states, "Phosphorus (P) is a critical nutrient for plant growth, and therefore food production for humans. However, there is an emerging concern that current practices are not sustainable for the long term due to increasing geological and geopolitical uncertainties about the supplies of P for fertilizer production and increasing demand for this nutrient due to population growth, growing affluence, and bioenergy production. There are currently no international organizations, policies, or regulatory frameworks governing global P resources for food security" (*Sustainable P*).
4. Units of organism populations.

WORKS CITED

"Apple App Store, iPhone 4, Angry Birds earn Guinness World Records." Technology. *Los Angeles Times*. 13 May 2011. Web. 9 Sept. 2012.

Billick, Ian, and Mary V. Price, eds. *The Ecology of Place: Contributions of Place-Based Research to Ecological Understanding*. Chicago: U of Chicago P., 2010. Print.

Cantrill, James G. "The Environmental Self and a Sense of Place: Communication Foundations for Regional Ecosystems Management." *Journal of Applied Communications Research* 26.3 (1998): 301–318. Web. 26 Aug. 2011.

Casey, Edward S. *The Fate of Place: A Philosophical History*. Berkley: U of California P., 1997. Print.

Christopherson, Harry Garretsen, and Ron Martin. "The World Is Not Flat; Putting Globalisation in Its Place." *Cambridge Journal of Regions, Economy and Society* 1.3 (2008): 343–349. Web. 13 July 2012.

Coppola, Nancy and Bill Karis, eds. *Technical Communication, Deliberative Rhetoric, and Environmental Discourse*. Stamford, CT: Ablex, 2000. Print.

Dobrin, Sidney I., and Christian R. Weisser. *Natural Discourse: Toward Ecocomposition*. Albany: State U of New York P, 2002. Print.

Donehower, Kim, Charlotte Hogg, and Eileen E. Schell. *Rural Literacies*. Carbondale: Southern Illinois UP, 2007. Print.

———, eds. *Reclaiming the Rural: Essays on Literacy, Rhetoric, and Pedagogy*. Carbondale: Southern Illinois UP, 2012. Print.

Goggin, Peter, ed. *Rhetorics, Literacies, and Narratives of Sustainability*. New York: Routledge, 2009. Print.

Herndl, Carl G., and Stuart C. Brown, eds. *Green Culture: Environmental Rhetoric in Contemporary America*. Madison: U of Wisconsin P, 1996. Print.

Killingsworth, M. Jimmie, and Jacqueline S. Palmer. *Ecospeak: Rhetoric and Environmental Politics in America*. Carbondale: Southern Illinois UP, 1992. Print.

Massey, Doreen. *Space, Place, and Gender*. Minneapolis: U of Minnesota P, 1994. Print.

Muir, Star A., and Thomas L. Veenendall, eds. *Earthtalk: Communication Empowerment for Environmental Action*. Westport, CT: Praeger, 1966. Print.

Orr, David. *Ecological Literacy: Education and the Transition to a Postmodern World*. Albany: State U of New York P, 1992. Print.

Owens, Derek. *Composition and Sustainability: Teaching for a Threatened Generation*. Urbana, IL: NCTE, 2001. Print.

Peterson, Tarla Rai. *Sharing the Earth: The Rhetoric of Sustainable Development*. Columbia: U of South Carolina P, 1997. Print.

Rice, Jenny. "From Architectonic to Tectonics: Introducing Regional Rhetorics." *Rhetoric Society Quarterly* 42.1 (2012): 201–213. Print.

Robinson, Kim Stanley. *Red Mars*. New York: Bantam, 1993. Print.

Said, Sally E. "From Oral Tradition to Legal Documents: Words to Protect the Headwaters of the San Antonio River." *Rhetorics, Literacies, and Narratives of Sustainability*. Ed. Peter Goggin, New York: Routledge, 2009. 116–131. Print.

Scialdone-Kimberley, H., and Metzger, D. "Writing in the Third Space From the Sun: A Pentadic Analysis of Discussion Papers Written for the 7th Session of the UN Forum on Forests (April 16–27, 2007)." *Rhetorics, Literacies, and Narratives of Sustainability*. Ed. Peter Goggin, New York: Routledge, 2009. 39–54. Print.

Semken, Steven. "Sense of Place and Place-Based Introductory Geoscience Teaching for American Indian and Alaska Native Undergraduates." *Journal of Geoscience Education* 53 (2005): 149–157. Web. 23 July 2012.

Sheller, Mimi. *Consuming the Caribbean: From Arawaks to Zombies*. New York: Routledge, 2003. Print.

Stein, Gertrude. *Everybody's Autobiography*. New York: Random House, 1937. Print.

Sustainable P Initiative. ASU Web, n.d. Web. 9 Sept. 2012.

Thompson, Krista A. *An Eye for the Tropics: Tourism, Photography, and Framing the Caribbean Picturesque*. Durham, NC: Duke UP, 2006. Print.

Tsing, Anna Lowenhaupt. *Friction: An Ethnography of Global Connection*. Princeton, NJ: Princeton UP, 2005. Print.

Urry, John. *Global Complexity*. Cambridge: Polity, 2003. Print.

Part I
Places We Dig (Mine)

1 A Certain Uncertainty
Drilling Into the Rhetoric of Marcellus Shale Natural Gas Development

James Guignard

Since 2007, there have been over 2,300 natural gas wells drilled in Pennsylvania. The effects of the drilling manifest themselves in many ways: increased diesel truck traffic, packed motels and rental houses, "man camps" scattered throughout the countryside, miles of insulated wire rolled out for seismic testing, and an influx of temporary offices, water storage tanks, and out-of-state license plates. This activity takes place above the Marcellus Shale, a band of natural gas-laden shale that stretches from West Virginia through eastern Ohio, western and north-central Pennsylvania, and into upstate New York. In 2008, Terry Engelder, a geoscience professor at Penn State, and Gary Lash, a geology professor at SUNY-Fredonia, estimated that the shale contained 500 trillion cubic feet of natural gas, about ten percent of which is recoverable. According to Geology.com, that volume of gas is "enough to supply the entire United States for about two years." Although the number has been disputed, notably by the USGS, there is enough gas in the Marcellus to attract many energy companies, including Shell, Exxon/XTO, Range Resources, and Chesapeake Energy. This gas rush is accompanied by a large, ongoing, multimedia conversation involving the industry, government officials, lobbyists, environmental groups, and the people who live over the shale. These agents' rhetoric shapes the perceptions of the Marcellus Shale region through numerous acts involving websites, press releases, TV, radio, and print advertisements, letters, contracts, phone messages, billboards and signs, news coverage, blogs, art (quilts, photography, pottery, and pastel landscapes, to name a few), bumper stickers, protests, books, bills, policy statements, scientific research, Google Groups, and public meetings. This ongoing conversation constitutes a corner of Burke's Barnyard *par excellence*. For this chapter, I drill into the rhetoric of the Marcellus Shale development, particularly in north-central Pennsylvania, where I live, although the conversation resonates nationally and globally. Because I live in the midst of an extractive process, I am most interested in the ways in which the industry and the locals who live here shape attitudes toward north-central Pennsylvania and how their rhetorical acts incline people to see this place.[1]

Gas drilling has become a mainstay of everyday life in Pennsylvania, and it promises to intensify, especially if the proposed 360,000 wells are drilled in 93,000 acres.[2] Nationally, natural gas is often touted as a bridge fuel to an alternative energy future and as a means to reduce our dependence on foreign oil. But this national public discourse hides the effects that occur on the ground when an industry moves into a region and begins extracting the gas. My research explores the competing rhetorics of the natural gas industry and local communities. In 2007, the industry began drilling and hydrofracturing the Marcellus Shale, a band of shale 5,000 to 8,000 feet below the surface. High-volume slick water hydrofracturing is a new extraction technology that involves pumping millions of gallons of water mixed with sand and chemicals into the drill bore and blasting the fluid into the shale. The liquid drains from the shale, leaving the sand behind to prop open the cracks, enabling the gas to flow into the well bore. Changes in Pennsylvania have been rapid, a sort of industrial *blitzkrieg*, and have left local communities reeling as they struggle to cope.

I argue that the industry uses a nationalized, displacing rhetoric that abstracts the region to ensure that gas development continues unabated, whereas communities use a localized, emplacing rhetoric that seeks to preserve a way of life by giving a face to the place. Although by no means symmetrical or balanced, the competing rhetorics present idealized versions of the worlds they inhabit and seek to create different perceptions of the region. For example, the gas industry ignores local knowledge in favor of their own language and practices (renaming roads with their own signs), whereas local communities discuss the need to preserve the small town feel and history of the area (overlooking the region's history with extractive industries). These competing views of north-central Pennsylvania lead, for the industry, to rhetorical strategies that create a geographical "blank" space useful primarily for the gas it provides and that moves Americans toward a "better future" and "energy independence." In response, local communities use rhetorical strategies to create an alternative sense of Tioga and the surrounding counties and their residents, one that asserts "We live here!"

The development of the Marcellus Shale has created what Marilyn Cooper calls a "web" of language (370), one that is connected intimately to the geography of the Marcellus Shale region. The words and images resemble the industry itself, with its well pads, roads, businesses, and pipelines interconnected across the land in order to move the gas. This web evolves as time passes, stretching, deforming, and reforming into new shapes and relations as each new rhetorical act emerges and works to shape attitudes toward whatever condition (unfettered access to gas; untrammeled public land; uninterrupted quiet at home) the agent desires. As Jenny Edbauer describes it, "we might . . . say that [the] rhetorical situation is better conceptualized as a mixture of processes and encounters" (13). Like the Marcellus Shale itself, the attitudes and emotions that constitute much of what people understand about natural gas development runs deep, spreading in all directions.

There have been millions of words written about the Marcellus Shale. Between 2000 and 2009, *The New York Times* database shows thirty-five articles for the phrase "Marcellus Shale"; from 2010 until this writing, the database shows seventy-three records. For 1,521 newspapers across the U.S., NewsBank shows one hit for "Marcellus Shale" in 2007; 216 hits in 2008; 403 hits in 2009; 2,014 hits in 2010; and 2,954 hits in 2011. Drawing on 2000 publications, the ABI/INFORM Trade and Industry database shows twelve hits for "Marcellus Shale" in 2006; eighty-five hits in 2007; 1,085 hits in 2008; 1,200 hits in 2009; 2,058 hits in 2010; and 2,311 hits in 2011. Significantly, ABI/INFORM shows how the industry's attention on the Marcellus Shale preceded the public's. I focus on salient examples of industry and local rhetoric to demonstrate the major strategies each group uses to argue for its version of the way this region should be perceived.

BIG MAN

Over the past three years, the rhetorical strategies used by the industry have evolved in response to a local public generally thrilled at first by the promise of an economic boom, then angered by the material effects of traffic, noise, and the threat of pollution. While the industry's rhetoric has evolved, several key strategies that shape attitudes toward the region nationally and locally have emerged. One key feature of these strategies is that they transcend their immediate geographical region, creating an impression of north-central Pennsylvania as open and implying that the region is abstract space (Tuan 6). Contrasting place with space, Tuan writes that "[p]lace is security, space is freedom: we are attached to the one and long for the other" (1). Tuan characterizes space as "a resource that yields wealth and power when properly exploited. It is a worldwide symbol of prestige. The 'big man' occupies and has access to more space than lesser beings" (58). Tuan's theorizing provides the framework needed to understand how the industry's presentation of north-central Pennsylvania as a frontier ripe for exploration recognizes the psychological importance for people to see resource-rich areas as such. The industry uses language that appeals to this sense of freedom and downplays the sense of place experienced by people who live over the gas. One example includes the renaming of roads from, say, Ore Bed Rd. to Pipeline Road 7, as happened to the road I live on. Rather than use the name that suggests knowledge of the place (Ore Bed), the gas industry uses a generic name that would work just as well in Texas, Wyoming, or Louisiana.[3] Such signs encourage whoever reads it to disassociate the resource from the place, thus downplaying the environmental impacts on people who live above the Marcellus Shale and who deal with the industry daily.

If there is a center for messaging, it's probably the Marcellus Shale Coalition, formerly the Marcellus Shale Committee. The Marcellus Shale

Coalition (MSC) is an industry trade group that features prominently in the news, especially MSC's president, Kathryn Klaber. Although there are other groups representing the industry, such as Energy In Depth, MSC is most prominent in the region. MSC acts as a focal point for much of the messaging about the industry, and rhetorical strategies used by MSC reverberate through the larger conversation about gas development.

We can observe several strategies summarized in the statement that greets readers on MSC's homepage. Only 129 words, the statement captures concisely the industry's major arguments for the development of the Marcellus Shale:

> The Marcellus Shale: Energy to fuel our future
> Welcome to the Marcellus Shale Coalition's (MSC) website. You'll find information on the Marcellus Shale formation, how we extract the natural gas and protect the environment, why we value the communities where we do business, and the opportunities that the Commonwealth and its residents can realize in the coming years and decades through natural gas exploration and production.
> You'll also learn about the important issues being addressed by our Marcellus Shale Coalition and the positive impacts natural gas drilling is already having on families, businesses and communities in many parts of Pennsylvania.
> So for some quick facts or more in-depth information, we invite you to explore the MSC website and discover why Marcellus Shale is the energy to fuel our future. ("The Marcellus Shale")

It's no surprise that MSC offers information on the process of extraction, which is clear and fascinating. However, the website also raises other issues, like "protect[ing] the environment," the effects of development on communities, and the promise of opportunities to families and businesses, which fall outside their expertise and are not as settled as MSC's language makes them out to be. But that's not surprising because the statement notes that the site is dedicated to sharing the "positive impacts" of natural gas development, a purpose that itself abstracts the region. Few changes on this scale are only positive.

Notably, the statement begins and ends with the phrase "energy to fuel our future." This phrase appears in differing versions in different contexts, ranging from politicians' utterances to newspaper editorials to industry CEOs' statements. For instance, responding to protestors outside an industry conference in Philadelphia, Chesapeake's CEO Aubrey McClendon stated, "What a glorious vision of the future [the protestors hold]: It's cold, it's dark and we're all hungry" (Rubinkam). Speaking about natural gas development at a manufacturing plant visit in Johnstown, Pennsylvania Governor Tom Corbett told a crowd of business owners and politicians, "It's not just jobs . . . It's national security. It's national defense. It's a future for our children, our grandchildren" (Siwy). McClendon conjures images of the Dark Ages, of regressing, which stands out in sharp contrast to a warm, bright, comfortable future his words suggest. Corbett raises issues of jobs, safety, and energy independence and ends by implying a promising, safe future for two generations of children. Such appeals to the future and safety echo throughout industry rhetoric, conjuring images of the frontier and taming the wilderness. As Lakoff and Johnson remind us, the future suggests a sense of progress, of moving forward or up (22), and it encourages people to look ahead, not at the present or past. Although the development of the Marcellus Shale holds promise, the industry's emphasis on the future turns attention away from the present and Pennsylvania's past experience with extractive industries. And because the future is an abstract fiction, the industry can choose the future they present. This framing aligns well with Tuan's notion of abstract space, and it can apply anywhere gas development occurs.

MSC's vision of the future focuses on economic development and jobs. This is particularly powerful at a time when, nationally, the economic outlook is bleak, and, locally, much of the drilling is occurring in rural areas that have lost manufacturing jobs and financial support for farming.[4] MSC addresses such concerns, claiming that "[t]he development of natural gas from Marcellus Shale offers great potential for the region's economic future, as well as the thousands of individuals, families and small, locally-owned businesses involved in extracting this clean-burning and abundant energy source from the ground" ("Opportunity"). Combining economics and the future, MSC states that gas development will benefit "thousands of individuals, families and small, locally-owned businesses," a characterization that encourages people to think about what may be, not what is. They support their claims with a Penn State University study called "An Emerging Giant: Prospects and Economic Impacts of Developing the Marcellus Shale Natural Gas Play." The study claims that "developing these natural gas reserves will directly generate thousands of high-paying jobs and indirectly create many others as employment is stimulated in support industries and as workers spend these wages and households spend royalty income" ("Opportunity"). MSC provides some numbers from the study's projected impacts, like jobs created and dollars generated, although they do not provide a link to the study.[5] MSC expects the appeal to expertise suggested by the academic affiliation to suffice for framing Marcellus Shale development as economically beneficial.

PICK A FUTURE, ANY FUTURE

MSC's emphasis on the future, economics, and jobs interjects possibility, security, and hope into the conversation about natural gas development by directing the audience to look to the future, specifically, the economic future, which discourages people from thinking about the present or a specific place. Looked at through Tuan's framework, the future is a psychological concept akin to abstract space—we may not know what the actual future will be, but we need such a vision in order to achieve a sense of progress. This economic framing serves the gas industry's purposes by projecting their vision of the future on this space, which appeals broadly to a national audience.

MSC claims that the industry works to protect the environment as well. Less than fifteen years old, slick water hydrofracturing is a major concern for many Pennsylvania residents because of the amount of water required, the amount of chemicals used, and the fact the technology outpaces the science. Yet the industry seeks to assuage public fears by portraying the environmental concerns as unwarranted. Industry rhetoric downplays the concerns by abstracting and simplifying the process, making the amounts of water used sound like a non-issue and creating analogies that link the water and chemical use with everyday items.

At a public informational meeting hosted by the MSC in June 2009, Pennsylvania Independent Oil and Gas Association (PIOGA) Executive Director Louis D'Amico stated that the natural gas industry will use the equivalent of one percent of the entire daily amount of water used in the state of Pennsylvania. He noted that the industry uses less water in a day than is required by Pennsylvania's golf courses. He also claimed that the chemicals used for hydrofracturing can be found in "consumer goods and cosmetics," which is true (Marcellus Shale Informational). D'Amico's choice of wording, however, overlooks significant differences between consumer goods under the sink or cosmetics on a shelf. For one, the scale is not comparable. A fraction of a percentage in a million gallons of water equals thousands of gallons of chemicals. Second, the public generally does not pump thousands of gallons of chemicals into the ground where they could potentially affect the environment. Instead, his phrase serves to abstract the chemicals in ways that put them into an innocuous framework by suggesting they are in our homes and used every day. His language suggests a just-more-of-the-same approach, which implies that the industry knows what it is doing, has been doing it for years, and is certain that the economic future they project will be untainted by pollution. So you shouldn't object now if you haven't objected before.

BIG BABY

Another way the industry rhetoric abstracts the region is by presenting contradictory frames depending upon context, a strategy that works well when projecting a particular vision for the future. Two framing metaphors the industry uses are industry as infant and industry as expert.

The gas industry often portrays itself in the stages of infancy whenever severance taxes are suggested. Severance taxes are an extraction tax, a tax applied any time a resource is removed from the ground. Pennsylvania is the only state involved in natural gas extraction without a severance tax, and the industry shapes their discussions of the tax in ways that suggest it would effectively stop resource development. The following press release is representative of the industry's rhetoric:

> Pennsylvania is blessed with rich natural resources, including a potentially large natural gas field in the Marcellus Shale. Although the MSC strongly opposes a broad-based severance tax, especially while the development of the Marcellus Shale is in its infancy, the industry remains willing to work through the Commonwealth's current financial challenges with the Governor and the legislature. ("Marcellus Shale Committee")

Four months later, Stephen Rhoads, President of the Pennsylvania Oil and Gas Association, writes in an op-ed, "Pennsylvania's Marcellus Shale industry is still in its infancy. [Pennsylvanians] seem to have the advantage in the current market for investment in shale gas development, and we think it makes good policy sense to keep the balance tipped in our favor." Using the metaphor of infancy in conjunction with discussions of severance taxes suggests the tax could crush a growing industry, costing Pennsylvanians jobs and opportunities and the nation a much-needed resource. In other words, a tax threatens a bright future. The choice also suggests that the public and politicians should take an indulgent stance toward an industry that has not yet learned to walk. You don't charge a child for the work he does. You give him an allowance.

When the industry discusses environmental impacts, however, they portray themselves as experts. A July 10, 2009, MSC press release on the proposed FRAC Act states, "Hydraulic fracture stimulation has been used in Pennsylvania since 1949 . . . Unnecessary regulation of this practice would hurt our nation's energy security and threaten our economy, and it would destroy Pennsylvania's shallow gas industry" ("PA's Oil"). Not only does this quote suggest that natural gas development has a long history in Pennsylvania; it links regulation to pain and suffering for the industry, plays upon fear, and downplays the fact that slick water hydrofracturing as practiced in Pennsylvania is less that fifteen years old. In another example, Penn State professor of petroleum and natural-gas engineering Robert W. Watson states that "one thing to keep in mind is that oil and natural-gas development is nothing new to the commonwealth," and hydraulic fracturing has its roots in "Pithole City, Pa., circa 1865, and Gulf Oil Co.'s laboratory research of the 1950s." Watson concludes, "But such development is not new. It is not unproven. It is not unsafe." Both men use unqualified language to alleviate public concerns by arguing that the technology has been used in Pennsylvania for years. The contrast with the metaphor of industry as infant is stark. The metaphor of infant is usually used in conjunction with severance taxes;

the metaphor of expert is usually used in conjunction with environmental regulation. Clearly, the gas industry wants it both ways, that is, they want to be left alone financially because they are a young industry muddling through infancy and regulation-wise because they are experienced.

The gas industry presents themselves as the experts on extracting natural gas, and rightfully so. However, much of their rhetoric shapes the public's perceptions in ways that presents their information as certain, even obvious, when much of it is questioned by disinterested parties. This is not surprising, given that the industry is trying to shape their rhetoric to meet two audiences—a national one and a local one. Their rhetoric primarily meets their needs nationally by creating a sense of the region as abstract, a frontier upon which the nation's projected rosy economic future depends. The industry's abstracting rhetoric overwhelmed local discourse during the first two or so years of their work here until their presence grew big enough to be widely disruptive. But the industry cannot just talk to the nation; they also have to talk to people who live here. Circumstances on the ground encouraged people to respond in whatever ways they knew, via letters, blogs, and the like. And although the residents of the region are split in their perceptions of the gas industry, a collective voice has emerged that offers alternative futures to the economic future painted by the industry.

WE ARE HERE! WE ARE HERE! WE ARE HERE![6]

In *Publics and Counterpublics*, Michael Warner explains that in order for a public or counterpublic to be created, it must be written into existence (16). The industry wrote one version of a public into existence, a conception of a public defined primarily by a bright economic future. Locals concerned about gas development encountered this frame when they sought information that took into account alternative visions, visions based on clean air and water and a quiet, rural life. Not all locals question the industry, but those who do find conflicting or contradictory information that undermines the industry's rosy projections. Made up of environmentalists, disgruntled landowners, and others, these locals began telling their stories of another future, hoping that their stories would disrupt the ones told by the industry. Although they do not have the PR professionals or funds enjoyed by the industry, the locals' identity as a counterpublic challenges the gas industry's notion that north-central Pennsylvania is simply a resource to be exploited for economic gain. Providing an important imaginative grounding, stories told by locals to locals and outsiders inhabit the region with flesh-and-blood humans invested in the place.

Like the industry, north-central Pennsylvanians have used multiple media to create a sense of themselves as a public and begin pushing back against the industry's vision for the future. Coming from environmentalists and disenchanted landowners, the locals' message grows from its rootedness in the soil, and it competes with the message of locals who for various reasons support the industry. One counterpublic grew from a Google Group called Citizens Concerned about Natural Gas Drilling (CCNGD). CCNGD characterizes itself as "nonpartisan group of citizens concerned with the potential negative social, environmental, and economic impacts of natural gas drilling development in Tioga County, PA." Open to anyone who cares to join, the group was created in October 2009 and has over 200 members. In 2011, members posted over 2,800 messages. Primarily, the group's purpose is to keep people abreast of the latest information about gas development, upcoming public meetings, plans for protests, deadlines for public comment, and the like. Representative subjects of posts include "Dumping well site soil behind Wysox Kiln Mill along river bank," "Urge EPA to Increase Public Participation in Natural Gas Decisions," and "How Natural Gas Development is Affecting Forestland [sic]." Invaluable for a potential counterpublic, such information provides a common knowledge base from which to question natural gas development, and it links strangers, an important component of a "public" (Warner 74), by giving them a common focus and reassuring them that other individuals share similar concerns. In other words, CCNGD creates a collective identity for the north-central Pennsylvanians who oppose an unchecked industry, an opposition that, once written into existence, empowers them.

CCNGD members recognize that the gas industry controls the message about gas development to a certain extent—one subject heading states, "40% of state drilling regulators have energy ties"—and they recognize the need to interject into the conversation terms that reflect their own concerns. Every so often, CCNGD members have exchanges that are about language or protest strategy that reflect their desire to challenge the industry's framing. One member recounted a conversation with an industry official who called her an "activist." Although she replied to the official that she was a "community member working to save our water and air," her e-mail to the group lamented, "I wish I had said if I am a water/air activist then you are an industry activist, a pollution activist" (12/14/2011). This led another member to express enthusiasm for the neologism, exclaiming "FABULOUS! I love the coinage 'gas industry activist' as a comeback, just as I love 'rig hugger' as the antonym of 'tree hugger' and 'pollutionist' as the counter to 'environmentalist'" (12/14/2011). This is one example of how a shared discourse can work to establish a counterpublic that then articulates their concerns and challenges the industry to respond. Whereas most messages are informative, the act of posting on and belonging to a Google Group that questions an industry creates a public that recognizes how the place and individual lives are affected. As Warner writes, "The peculiar character of a public is that it is a space of discourse organized by discourse. It is self-creating and self-organized; and herein lies its power, as well as its elusive strangeness" (68). The CCNGD Google Group provides a virtual space for people who live here to share their information and their stories, thus serving as an electronic proxy for the physical region. It creates a public that bonds readers and reminds them every time they log on that they are not alone, that people who share their concerns and values live nearby. In this way, the Google Group works to counteract the abstracting rhetoric of the gas industry.

CCNGD exists primarily for the benefit of its members, and part of its function is to create a shared sense of place. However, it is indicative of the way that the people of the region coalesce into a public that then serves to support rhetorical acts directed to the larger public that contains it. These rhetorical acts exist in many media—letters to editors or politicians, blogs, protests, and the like. One rhetorical strategy that many of these acts share is the attempt to populate the region with people rather than framing it in abstract economic or environmental terms. Several regional blogs work to do this in varying ways: Ada Mae Compton, The Faces of Frackland, and Honesdale Concerned Citizens. The bloggers who administer these blogs question the gas industry's message from the position of a person emplaced in the region, which invites connection and shared history. Ada Mae Compton is the name of a three-legged rescue dog and the voice of the blog ("My owner says . . ."). The Faces of Frackland consists of pictures of individuals accompanied by their experiences with gas development. Honesdale Concerned Citizens focuses primarily on issues of zoning in Bradford County, Pennsylvania. That's not to say the north-central Pennsylvania blogs do not use academic and scientific reports to question the industry. They do. But the bloggers, like the members of CCNGD, intertwine personal stories and accounts with their appeals to expertise. They invite the "pause in movement" Tuan claims is so crucial to creating a sense of place (138). Once this pause in movement occurs, the momentum of the industry is blunted, because the audience is invited to pause and to think of north-central Pennsylvania as something other than a resource.

A CERTAIN UNCERTAINTY

There are many factors affecting the development of the Marcellus Shale, but the growth of a counterpublic that questions the industry and the emergence of a rhetoric that puts a face on the place contribute to the growing awareness of the people who live here and those who don't. Google Groups, blogs, letters, and protests do not equal a place, but they do provide the discourse necessary for creating an image of the region as more than a resource to be developed. And although seeing the region as a place does not guarantee that the industry will become more careful or sensitive while extracting the gas, it does ensure that other parts of the story are being told.

By presenting their reasons for natural gas development in absolute terms, the industry creates the sense that it is an unimpeachable source, even though much of what it addresses is outside its expertise and far from settled. Penn State professor Terry Engelder sums up the industry's attitude when he speaks to the need for Pennsylvanians to "sacrifice" in order to realize the positive outcomes of the Marcellus Shale development for all Americans (Adelbeck). Although Engelder does not speak for the industry, his characterization of the situation as one requiring sacrifice recognizes that people live here while implicitly supporting the industry's framing of the region as free for exploration.

But the locals' rhetoric appears to be vibrating through the natural gas industry's web in meaningful ways. Recently, Chesapeake Energy began their "This Is Our Home" ad campaign across Pennsylvania. Full-page newspaper ads proclaim "This Is Our Home" above a picture of a drill rig in the Pennsylvania landscape, and followed by text that echoes MSC's message: Chesapeake "take[s] pride in safe, environmentally friendly operations as we create energy security for America, revenue for local communities and jobs for Pennsylvanians." Below the text, the ad displays thirty-six 1.2" x 1.3" portraits of individuals who work for Chesapeake, including their names, job title, and where they presumably live. The ad lends the Oklahoma-based Chesapeake a sense of rootedness in the region and recognizes the company's desire to align itself with local concerns. Chesapeake's ad shows the industry attempting to shape specific discourse for local audiences, something that hasn't occurred until recently.

The failure to articulate a rhetoric that meets local and national concerns has created problems for the industry. There are no doubt many reasons for this mistrust, but part of the problem stems from the industry's persona as too damn sure of itself. After a while, this comes across as smug, and if questions go unanswered, it becomes condescending. *Y'all just sit back and let us do our jobs. We know exactly what we're doing.* Stories told by groups like CCNGD ask audiences to consider alternative futures, futures that industry rhetoric does not account for. Instead, the industry responses are vague and contradictory, and lead to more confusion. I'm convinced that's what the responses are designed to do. As long as it's the counterpublic's responsibility to provide facts and data, the industry keeps drilling at top speed. But even politicians at the national level are concerned about the growing mistrust. President Obama has asked the Department of Energy to assess slick water hydrofracturing's safety more quickly than the Environmental Protection Agency's study timeline calls for (Lustgarten). The example of public mistrust derailing the nuclear industry is raised as a warning. A recent study from Michigan shows that although Pennsylvanians expect more positive impacts than negative ones from gas development, they are nonetheless concerned about water pollution and the industry's close ties with Governor Tom Corbett (Rabe 2).

It's too late to turn back the industry's push for an economic future, but it's not too late to talk about—and work toward—alternative futures. One way involves the gas industry (or any extractive industry) defining the future in terms broader than economic. Another change involves providing an honest accounting of what is and is not known about the industry's development, not only the "positive impacts," as MSC states. The stakes are too high for the people who live here to be persuaded by a vision of the future that is only economic and fairly short term—twenty, thirty, forty years. Counterpublics have emerged with alternative visions of the future, writing long-term futures that take into account the ground and air and water where they live, and they have shifted the conversation locally and nationally. Perhaps we will find that the economic future can coexist with other futures, perhaps not. But the industry and the nation now recognize that the drilling is occurring in a place, and that recognition shifts the perception of the land in which the industry drills.

NOTES

1. Full disclosure: I am a landowner who has chosen not to lease. I am not opposed to drilling, but I am deeply concerned about the impacts of the industry. My wife and I blog about our experiences living here at Greetings from Pipeline Road 7, and I am a member of Citizens Concerned about Natural Gas Drilling.
2. Although the numbers surrounding the development of natural gas constitute a large part of the rhetoric used, I will not be focusing on them in this analysis. They have been revised multiple times in the last three years and are in constant dispute even now.
3. There are likely several reasons for this: out-of-state workers don't have time to learn original names; safety vehicles need greater detail about pipeline sections. However, there's no denying the rhetorical dimension of the signs.
4. This has changed since the gas industry began work. According to the Bureau of Labor Statistics, Tioga County has one of the lowest rates of unemployment in Pennsylvania. See http://www.bls.gov/ro3/palaus.htm.
5. The report can be accessed at: http://s3.amazonaws.com/propublica/assets/monongahela/EconomicImpactsMarcellus.pdf.
6. Remember Dr. Seuss's *Horton Hears a Who?*

WORKS CITED

Adelbeck, Hannah. "The Risks of Marcellus Shale Drilling Are Worth the Potential Gains, Says PSU Prof." *Voices of Central Pennsylvania*. 10 June 2010. Web. 9 Jan. 2012.
Citizens Concerned about Natural Gas Drilling. Google Groups. Oct. 2009. Web. 13 Feb. 2012.
Cooper, Marilyn. "The Ecology of Writing." *College English* 48.4 (1986): 364–375. Print.
Edbauer, Jenny. "Unframing Models of Public Distribution: From Rhetorical Situation to Rhetorical Ecologies." *Rhetoric Society Quarterly* 35.4 (Fall 2005): 5–24. Print.
Lakoff, George, and Mark Johnson. *Metaphors We Live By*. Chicago: U of Chicago P, 1981. Print.
Lustgarten, Abrahm. "Fracking Cracks the Public Consciousness in 2011." ProPublica. 29 Dec. 2011. Web. 6 Feb. 2012.

"Marcellus Shale—Appalachian Basin Natural Gas Play." Geology.com., n.d. Web. 2 Jan. 2012.

"Marcellus Shale Committee Issues Statement on Severance Tax." Marcellus Shale Coalition, n.p. 4 Feb. 2009. Web. 26 June 2009.

Marcellus Shale Informational Meeting. Marcellus Shale Committee [now Coalition]. Wellsboro High School. Wellsboro, PA. 4 June 2009.

"Opportunity." Marcellus Shale Coalition, n.p., n.d. Web. 16 Jan. 2012.

"PA's Oil, Gas Industry Opposes Fracture Stimulation Reporting Bill in Congress." Marcellus Shale Coalition. N.p. 10 July 2009. Web. 20 Oct. 2009.

Rabe, Barry G., and Christopher Borick. "Fracking for Natural Gas: Public Opinion on State Policy Options." Center for Local, State, and Urban Policy. U of Michigan. Nov. 2011. Web. 2 Feb. 2012.

Reynolds, Nedra. *Geographies of Writing: Inhabiting Places and Encountering Difference*. Carbondale: Southern Illinois UP, 2004. Print.

Rhoads, Stephen W. "Marcellus Shale Tax Won't Solve Budget Woes." *Penn-live.com*. 15 June 2009. Web. 26 June 2009.

Rubinkam, Michael. "Gas CEO Takes on Protesters." *Times Leader*. 8 Sept. 2011: 4A. Web. 10 Oct. 2011.

Seuss, Dr. (Theodor Geisel). "Horton Hears a Who!" *Storytime*. New York: Random House, 1974. 66–124. Print.

Siwy, Bruce. "Corbett Touts Johnstown Manufacturing Plant." *Daily American* 22 Dec. 2011: N.p. Web. 29 Dec. 2011.

"The Marcellus Shale: Energy to Fuel Our Future." Marcellus Shale Coalition, n.p., n.d. Web. 16 Jan. 2012.

"This Is Our Home." Chesapeake Energy. *Wellsboro Gazette*. 21 Dec. 2011: 14-A. Print.

Tuan, Yi-Fu. *Space and Place: The Perspective of Experience*. Minneapolis: U of Minnesota P, 1977. Print.

Warner, Michael. *Publics and Counterpublics*. New York: Zone Books, 2002. Print.

Watson, Robert W. "Sunday Forum: The Bottom Line on Marcellus Shale." *Pittsburgh Post-Gazette*. 19 July 2009. Web. 10 Oct. 2009.

2 Eco-Seeing a Tradition of Colonization
Revealing Shadow Realities of Marcellus Drilling

Brian Cope

[After] loggers swept across the northern tier of the state more than a century ago [. . .], Pennsylvania gave away the store to the coal barons, too. They gouged hillsides, destroyed drinking water supplies, contaminated thousands of miles of streams, and left a cleanup tab in the billions of dollars. (Stranahan)

Stranahan's editorial represents a microcosm of what amounts to a tradition of colonization in rural Pennsylvania. In *Colonization in America: The Appalachian Case*, Helen Lewis describes a historically colonized Appalachia punctuated by outside interests who exploit resources and then blame the locals when things go wrong (2). This is a continuing pattern. The boom-and-bust economies that come with this tradition have altered perceptions and communities and caused author Tawni O'Dell to observe a "sense of deprivation" in her home region of western Pennsylvania. Considering the subject of this collection, this chapter contemplates how the rhetorical ecology of my home place interacts with a tradition of energy extraction that has brought deprivation, environmental devastation, and boom-and-bust economies to rural Pennsylvania hills, vales, and waterways. It flows from the rhetoric surrounding the current incarnation of this tradition, the process of hydraulic fracturing (herein "fracking") in western Pennsylvania[1].

Specifically, this chapter focuses upon commercials paid for by the corporation, Range Resources, in conjunction with the surrounding fracking-related milieu in Pennsylvania. Moving beyond the commercials' obvious green-washed images of healthy waterways and the pastoral idyllic, I aim to illuminate both an active concealment of health and ecologic concerns and a pervasive economic, industrial rhetoric shadowed in agendas of greed and cynicism. As noted, such shadows exist in a place where energy extraction has long been a colonizing tradition that has permeated and co-opted the rhetoric of my home region, obfuscating a more natural, holistic rhetoric, and subverting a proud tradition of working people. To heighten an understanding of various rhetorical complexities, I view this milieu through the lens of often false, industry-created exigencies as well as the term "reverence." Applying this lens to the context of the tradition of colonization, I examine the commercials alongside the rhetoric of politicians and industry, first, to convey how the commercials fuel a colonizing agenda of the fracking industry, and second, to illustrate how the awareness of false exigencies and irreverence can lead to invention that unveils a rhetoric of place that moves beyond destructive traditions and the rhetoric of negation to reflect the diversity and holism of my home place.

Sitting on my back porch in late spring, the ecological diversity of rural western Pennsylvania crystallizes in the many species of bird song. An amateur ornithologist has told me that in mid-summer over 150 species frequent her backyard avian oasis. A similar diversity can be found in the Appalachian flora, plant-life that Native Americans and the subsequent Appalachian peoples knew intimately in terms of healing and spirituality. Rhetoric can and does spring from these nonhuman elements of place, as many eloquent writings, such as Rick Bass's many defenses of Montana's Yaak valley, attest to. In rural western Pennsylvania, one senses this rhetoric seeping from hills and valleys into the meetings of groups such as the senior environmental corps composed of numerous retirees—many of whom used to work in the mining industry—who now devote their time to testing waterways for industry-created imbalances; or the many other volunteers who work tirelessly in local conservation and watershed organizations. After irresponsible mining practices turned many waterways orange with iron ore in the 1960s and 1970s, many now see this region as place of recovery. Their rhetoric is one of balance and holism that springs out of their action and what is right for the region.

However, despite the efforts of local environmental groups, the powerful agenda trumpeting energy extraction tends to drown out a holistic rhetoric of place in favor of a monochromatic rhetoric of deprivation focused solely on jobs and need; these powers benefit from the mindset of the tradition of colonization. However, the tradition of colonization often alters the public rhetoric of these groups in rural western Pennsylvania. Because of a pervasive fear of being seen as "too radical," they often begin at a middle-ground position. These middle-ground positions are then spun by industry and politicians to appear extreme. Thus, it is vital to understand the weight of the feeling that the tradition of colonization has bestowed upon the region and how it has contributed to the layered complexity represented in the rhetorical ecologies of the region.

Many feel that colonization hit full-swing with coal (Walls and Billings), and although not as devastating as in West Virginia, where companies blow up mountains, it is every bit as palpable. Numerous Appalachian scholars have noted that energy colonization began with the mining industry. After noting a number of contributing factors, Edward Abbey claims that the worst aspects of this colonization leading to deprivation were caused by the broad form deed that "not only gives the buyer of mineral rights the privilege of extracting the coal from the farmer's land, it also gave him the option to remove that coal 'by any means necessary'" (84). In such operations, mining companies could legally strip-mine a person's backyard, causing

Kentuckian Wendell Berry to describe the degraded land as being passed "from generation to generation like a family curse, a birthright of ugliness and diminished hope" (28). Adding to the diminished hope in this region are the detritus of industry—the bony piles, coal tipples, and orphaned gas wells. The rhetoric that springs from this degraded land through the tradition of colonization is beholden to economic promises often by outside companies of relatively temporal jobs. I noticed such deprivation when I first moved back to the region from the west. Despite the verdant richness and sometimes stunning beauty of this place, people seemed to want to be working so that their children could go to college and live elsewhere. A sixth-generation acquaintance on the local planning commission describes this colonization as a mental phenomenon, as the residents think of themselves as second-class citizens who feel the need to accept any offer of jobs no matter how the odious the industry. Thus, in the tradition of colonization, the rhetoric of this place becomes one of deprivation that is synonymous with rhetoric of energy extraction. The commercials exist within this context. Herein, I examine them as part of a systemic campaign to gloss over potential negative health, public policy, and infrastructural issues.

The commercials—played on Pennsylvania television stations and available on the *Myrangeresource.org* website—are one corporation's reaction to the hype and shock value of the film *Gasland*[2] and hundreds of allegations of fracking-related environmental problems. They build pathos by depicting the beauty of Pennsylvania, in conjunction with a steady and responsible corporate partner. Meant to reassure a population that has heard and seen many negatives about drilling, the commercials feature vignettes about various real-life, Pennsylvania residents, representing leaseholders, employees of Range, small business owners, and other community members. They borrow from a genre known as eco-cinema in that they have both "stunning" and "arresting" qualities, which Killingsworth and Palmer note could cause "the cessation and truncation of action" (301). These apathy-inducing qualities are derived through the portrayal of pristine-looking ecologic and iconographic images that Morey has termed "econs." In this chapter, one of these econs takes the form of a Pennsylvania waterway upon which a father and son are fishing; a second represents the aesthetics of the pastoral and a slumbering yet bucolic farming culture. Dobrin and Morey's concept of "eco-see" provides an entryway into the complexities of how the commercials operate. Along with being a way to analyze images, they envision eco-see in part as "how humans use images to construct ideas of nature and environment and how these images create and reinforce those constructions, and how humans may use existing images (or make new ones) to create alternative ways of seeing nature and environment" (8). The frame of exigency and reverence furthers these ideas to work toward the creation of such alternative images. Exigency has been defined by Bitzer and others as a necessary element of a rhetorical situation. I agree with this, of course, but wish, however, to take the analysis of the term into the direction of what I call false exigency, the kind that is fueled by a rhetoric that promises unsustainable jobs. Herein lies the problem for the people of this region: the tradition of colonization has caused economics of the place to involve the destruction and changing, and co-opting of the rhetoric of this place. However, to fully understand the rhetorical frames in the tradition of colonization as they relate to fracking, here's a bit of context.

Embedded in shale far below much of Northern Appalachia lies a repository of natural gas that industry people call the "Marcellus Shale play." Above ground, the idea of Marcellus Shale gas was an obscurity until two independent occurrences set in motion a gold rush. In 2007, Range Resources had great success in a "test" drilling operation in southwestern Pennsylvania, and Penn State Geologist Teri Engelder made the "mind-boggling" calculation that "50 trillion cubic feet of natural gas could be recovered from the Marcellus formation, or the same amount the entire country uses in about 21 1/2 years" (Silver). The process of accessing these deposits involves 5,000 to 9,000 feet of vertical drilling and up to 10,000 feet of horizontal or directional drilling. To free the embedded gas, companies shoot a mixture of water, sand, and chemicals (many of them toxic) to fracture the shale. Technology is such that gas companies can horizontally drill and fracture in several directions, steering the probes with precision (Bryzcki). The fracking process amounts to a major industrial operation, far more intensive than the previous shallow gas well drilling as it generally requires a few acres and roughly from one to five million of gallons of water depending upon how the large the operation.

Since the groundbreaking discoveries of both Range and Engelder, wading into the rhetorical swamp that is Marcellus shale fracking, a person can become mired past hip-boots in half-truths, horror-story anecdotes, conflicting studies, and varied perceptions of shock value. Environmental problems, fears, and allegations have mingled with corporate and lease-holder profits to form a contentious dialectic that ranges from a Palinesque "Drill, baby, drill!" to drilling bans and "Fracksylvania" bumper stickers. Urban-based groups such as MarcellusProtest.org call for bans whereas groups in my home region, such as the Coalition for a Healthy County, set more modest goals due to fears of appearing too radical. "Facts" unfold every day and often differ depending upon who is stating them. Industry-sponsored "grassroots" websites provide positive anecdotes antithetical to activist sites such as *MarcellusProtest.org*. Things would be much easier if the earth would issue a deafening scream when we have gone too far, as it did in when Arthur Conan Doyle's arrogant Professor Challenger drilled eight miles deep in "When the Earth Screamed." No such luck, though, and despite more subtle screams, manifesting in a gaggle of many public health and environmental problems, deep-well Marcellus fracking is being fed to Pennsylvanians by the industry and state political leadership as the economic opportunity of a lifetime (M. White).

In May of 2011, the head of Pennsylvania's statewide Marcellus Shale Task Force, the Lieutenant Governor Jim Cawley, told our county task force that there has "never" been a water well ruined by fracking in Pennsylvania, a common refrain by the industry. As part of a very industry-friendly administration, Governor Corbett having collected over one million dollars in campaign contributions from the gas industry during the last campaign (*Marcellus Money*), Cawley implored us to first make sure we get the facts and not listen to media and *Gasland* horror stories. Then he told a "joke" about the heavily drilled area around Dimmock in Bradford County, Pennsylvania, where *Gasland* Director, Josh Fox, filmed numerous residents setting their well water ablaze from sink taps. He said that residents in that area have been lighting their water on fire for decades and that the methane migrations that caused such tap-water blazes happened during the "shallow" and not the "deep" gas boom. He has made similar statements to at least one other venue as reported in *The Philly Burbs*, who noted that that the DEP has found that "improper well casing and cementing was blamed for gas contamination of 16 families' drinking water supplies" (McGinnis). It was not actually the fracturing of the rock but faulty cement casing that caused the contamination of these water supplies. As this casing is a vital part of the fracking process, Cawley was engaging in a duplicitous rhetorical game by drawing upon a rhetoric of deprivation in the colonizing tradition of energy extraction.

Beyond Cawley's manufactured rhetoric of deprivation, here is a summary of environmental issues associated with fracking. In 2011, the *New York Times* found that frac-flowback water near Indiana Pennsylvania, about a mile from my home, was contaminated with 200 times the legal limit of radium allowable in water supplies (J. White). At the time, this was a major concern because companies regularly practiced "diluting" this flowback into Pennsylvania rivers and streams, illustrating an ongoing problem in finding a way to safely dispose of this toxic brine. Moreover, geologists feel that injecting wastewater in deep-injection wells has caused unlikely earthquakes in Ohio, Colorado, Texas, and Arkansas (Phillips). A Duke University study concluded that methane levels were seventeen times higher around fracking operations (Osborn et al.), and despite natural gas being a "clean-burning" fuel, a Cornell study found that the cradle-to-grave greenhouse gas footprint of fracking operations has the potential to add more carbon dioxide into the atmosphere than coal (Howarth et al. 688). Another study suggests a possible link between the disposal of fracking wastewater with the deaths of livestock and house pets (Di Paola). Aside from these examples there have been hundreds of allegations by private residents, which companies in general deny. Problems are such that the editors of *Scientific American* have warned that states like Pennsylvania are "flying blind" and need to slow down gas production to allow the technology and regulations to keep pace with the drilling ("Safety First").

Cut to the Range Resource commercials. A steady, melodic, acoustic folk riff plays while the viewer takes in alternative views of a forest, home, or vast fields of green. Substitute images of healthy cattle, or children, or a couple walking hand-in-hand—lots of sunshine and white puffy clouds. In most of the videos a man, perhaps the owner of a farm or business, conveys images of the pastoral idyllic, speaking about nature or how he loves his dog/wife/cattle/farm/job, depending upon the particular story in the commercial. A few of the videos feature women, a single mother who has a job as a Range secretary, a grandmother leaseholder who is able to pay for her grandchildren's college expenses, whereas another owns a stable. The characters—real-life company employees, community members, business owners, and leaseholders—all express their gratitude to Range Resources (*MyRangeResources.com*). Through the totality of these videos, Range Resources presents a holistic effort to legitimize their presence in Pennsylvania, utilizing both the beauty of the region and the false exigencies of jobs created by a tradition of colonization.

Much is lost through false exigencies imposed upon us by powerful entities, masking hidden agendas. One illustration of the latter is contained within the *History of Madness*, wherein Foucault uncovers that the hidden agenda behind the sudden proliferation of insane asylums in the 1700s Europe was due as much to the curing of leprosy than the official reason. The root of the term "exigency," which the OED defines as an "urgent want; pressing state (of circumstances)" comes as a demand in law, a warrant occasioning "the sheriff to summon the defendant to appear and deliver up himself upon pain of outlawry." Those whose mechanized activities are fueled by greed often pursue these activities with the full force and command of laws written specifically to protect them. Our world abounds with destructive examples of these types of exigencies. For instance, it has been perfectly legal for mining companies to blow up mountains in the service of an "urgent" need for electricity, even though such a process destroys, or forever changes local watersheds and causes countless health problems. Pennsylvania Oil and Gas law is extremely powerful, so much so that local governments often cannot enact regulations to prevent mineral extraction in certain places due to the risk of lawsuit. Strengthening these regulations Pennsylvania recently enacted HB 1950, which establishes statewide regulation that strips a locality's ability to zone for fracking activities (Friedenberger). As stated, the message from politicians and the industry is that this is the opportunity of a lifetime, as if the gas wouldn't be there for those in the next lifetime. Thus, the exigency needs to be examined in terms of the context of how it serves a tradition, a colonization, and rhetoric of deprivation; an understanding of reverence works to clarify this context.

According to Paul Woodruff, reverence was heralded by Greek poets but downplayed as a virtue by Plato, perhaps due to Plato's concern that leaders would use it to abuse their power, causing the true meaning of reverence to also be deemphasized for centuries (9). However, as Woodruff shows in his work *On Reverence*, its ancient origins in the west and east have more to do as a check on power, rather than an opportunity to abuse it. Simply put, reverence works to "keep humans from acting like gods." Woodruff begins from a "schema" of the term he describes as the "well-developed capacity to have the feelings of awe, respect, and shame when those are the right feelings to have," giving examples of feeling awed by "a great whale, a majestic redwood, or a range of tall mountains" (9). He finds that reverence has more to do with power than religion. Yet, because people mistakenly equate reverence as something involved with blind ritual or dogma, Woodruff separates and defines various aspects of this virtue; this includes irreverence, which he notes is often misused in the place of boldness or subversion. In so doing, Woodruff describes a modern American society that has misconstrued the idea of irreverence, in that satirizing or criticizing a corrupted official or institution is held up as being irreverent. This could not be further from the original intention of the idea of reverence. Rather, this person's boldness may have been occasioned by irreverence on the part of the entity in power (36). From Woodruff's perspective, the bold, the creative or the subversive often act out of reverence.

This comparison of reverence and irreverence is not to reduce this complexity to a binary but to show how this shift can reveal the hidden agendas and complexity within power structures. In this case, such a lens exposes the irreverence of both Lieutenant Governor Cawley and Range Resources, evidencing a disingenuous rhetorical agenda contrary to the tone of the commercials. Cawley's insouciance about the families' documented contamination of water wells evidences irreverence toward both his constituency and home place. More perniciously, his rhetoric implies that the land has already been "ruined" by previous industries. Thus, an understanding of exigency and reverence moves beyond obvious green-washing to discover agendas and rhetoric concealed in the shadows of this situation. The effect of this campaign is to promote a rhetoric of conflicted and confusing images that work on the affective level.

Range Resources has also displayed such irreverence for this rural area. First, the company has taken part in what appears to be an industry-wide effort to discredit claims and deflect blame such as in a water well contamination in Texas. Such deflection has been common enough to be the subject of parody. For instance, when Range Resource representatives deflected blame for water well contamination as a result of Barnett Shale Drilling in Texas, an exasperated blogger wrote a satirical piece blaming the "fart fairies" not the gas companies for contaminating water wells (Southwell). Second, traveling away from the *MyRangeResources.com* site to the corporation's main page reveals the company's reverence clearly lies with the idea of making a profit and "cutting costs." The script on the website conveys that "Range Resources seeks to consistently drive up production and reserves, while maintaining one of the lowest cost structures in the industry in order to increase shareholder value." Nothing is mentioned about concern for leaseholders, communities, or the environment. Third, spokesperson Matt Pitzerella boasted at an industry conference that Range employs former military "psychological operatives" to work with residents and to help localities "write ordinances" in Pennsylvania (Javers). Supporting these points, ethnographer Simona Perry has found that the relationship between the fracking industry and the citizens of Bradford County resembles that of an abuse cycle. She gives a particularly Foucauldian example of a local fifth-generation farmer and former school board member who was incarcerated for protesting two separate spills on his land and committed for psychiatric evaluation, where he was diagnosed as having bipolar disorder. With this understanding of the false exigencies combined with the irreverence toward the people of Pennsylvania, one begins to understand the commercials as drawing from both econs to further a tradition of colonization that has led to a rhetoric of deprivation.

As evidenced above, identification of exigency and reverence represents a gaining of awareness that works beyond Range's green-washing. This comports with Morey's rhetorical discussion of eco-see and provides a malleable foundation for the viewing of the commercials. In "A Rhetorical Look at Ecosee," Morey carves a path for the viewing of images beyond hermeneutics, showing that writing is not just a "vehicle to this world but its ontology" (23). Morey conveys that images can also have such representative qualities; for instance, the image of a meadow full of wildflowers does not simply represent nature, but also composes it (24). Thus, eco-see employs a full range of awareness-creating approaches to the understanding and embodiment of ecological images. Morey offers a literary example, describing how Edward Abbey's characters in the *Monkey Wrench Gang* chopped down newly erected billboards in order to "'reoccupy' this space for what they call 'nature'" (43–44). The suggestion then is that we move beyond interpretation of the images into an invention that changes them. In terms of the commercials, the understanding of how reverence pairs with imposed exigencies can cut another path to help us gain such awareness.

The commercials' exploitation of pastoral econs has a powerful impact on the psyche of Pennsylvania residents as views of family farms punctuate the Pennsylvania countryside. Moreover, unlike the vast majority of homeowners in Pennsylvania, family farmers tend to own their mineral rights because in most cases the land has been in the family for generations. A number of commercials prey upon reverence to depict how Range has improved families and relationships. One of the more powerful contentions deals with the idea that Marcellus drilling has helped farmers to avoid foreclosure. For instance, the video "My Opportunity" commercial depicts an admirable older man who has been farming all of his life, having farmed at one point 1,000 acres when he was grain farming, and having worked various jobs in order to keep his farm. He describes how being a farmer is difficult because the prices they receive are the "same as forty years ago," whereas the price of machinery and other costs have skyrocketed. Standing beside his barn, he talks about how all the farmers in the area have gas well leases. After a few moments, pathos builds as his wife joins him and reveals that they met in grade school, playfully noting that she has always known that the farm has always been his first love. She continues to describe how "natural gas has been a godsend to for the area" and that many farmers have used their royalty check to upgrade their farms.

The second econ merges fishing in Pennsylvania with the timelessness of a father and son relationship. Biologist Sandra Steingraber describes our powerful connection with water: "amniotic fluid is creeks and streams that fill rivers." In Pennsylvania, environmental consciousness often begins with fishing. Amid northern Appalachia's abundance of waterways, people tend to develop an affinity for particular places, the bend of river, a beaver dam, a soft bed of needles in the hemlocks that will be a resting place for the evening. Eventually, a person knows a place so well that it becomes part of her; out of this embodiment, reverence grows, and environmental concerns often blossom. While the father and son both tell their stories, the signature melodic folk song plays in the background, and the viewer takes in alternative images of father and son fishing, and tree-covered rolling hills. After graduating from a Pennsylvania university with a fisheries degree, the son had to take a job out of state and was only able to return after being offered a job from Range Resources. Range's ethos is bolstered in the idea that the son returns as a water quality expert to protect his home place, along with his other coworkers who, as he mentions, also fish and hunt, and thus care about the environment. Toward the end of the video, a silhouette shadow image of the father and son fishing is reflected in a lake. What can be found in the shadow images of these videos? And how can the understanding of these shadows lead us to a meaningful new place beyond the thesis, antithesis, synthesis of the traditional dialectic where all parties seem to be caught?

The recognition of such shadow images does not negate the videos. The emergent alternatives herein work beyond the "drill," "anti-drill," and middle-ground platforms to unveil larger truths. This is where Muckelbauer's ideas of affirmative "singular rhythms" become useful. He feels that most dialectics are limited to the practice of negation, noting that "[w]hen faced with the ethical and political dangers of dialectical change, the challenge is to invent a practical style of engagement that doesn't just repeat the structure of negation and refusal" (12). A singular rhythm represents something new that emerges from within the repetition of a negating dialectic. This affirmative style of reading "is not primarily concerned with extracting a thesis or claim (whether in order to explain, contest it, advocate it, or develop it)" (43).

Further illuminating the need to stray from a negating dialectic is the need to identify and transcend rural Pennsylvania's tradition of colonization by energy extraction. In Pennsylvania, this tradition has blinded us to certain realities. For instance, the "land is already ruined" mindset has been incorporated into the Chapter 93 of the Pennsylvania Code, which designates certain "high quality" streams as being protected from industry. Although this would seem an environmentally focused law, it assumes that the other waterways have already been "ruined." The question becomes, why should not all streams be "high-quality?" Was the lake where father and son fished high quality? Moreover, Pennsylvania has other indicators of the energy colonized, with outside investors owning much of the land as a Penn State University study has shown that about half the leaseholders are from out of state (Kelsey et al.). Thus, seeing mineral extraction as a tradition creates a certain amount of comfort and complacency; whereas recognizing the activity of this tradition in colonization tends once again to be a method of gaining awareness.

Building upon this awareness of the false reverence of Range, the irreverence of elected officials, the deflection of blame by the industry, and laws that designate only certain waterways as high quality, what lies in the shadows of each video? In the farm commercial, gas drilling is presented as an exigency allowing the couple to continue on their land. According to Eileen Schell, in the "Rhetorics of the Farm Crisis," this is the avoidance of a tragedy narrative due to a concept called "smart diversification." Too often, Schell notes, smart diversification stresses an individualism that fails to include the health of land and sustainable agriculture (84). Also, although the farmer in the commercial alludes to difficult economic issues of farming, what is not discussed is that the forces causing the family farm's demise are very similar to the forces governing our energy policies. Since World War II, federal policies have subsidized agribusiness to the detriment of the smaller farmer (84). Another factor that lies in the shadows is the growing interest by consumers in local sustainable farms. Schell mentions the group "Farm Aid" that supports "the image of the small farm . . . that is not merely nostaligic but an image of resistance to neolibralism and factory farming: a place that is rooted in family and community" (102). Also not discussed are the various water wells contaminated by the process of fracking. Localism and community can be an effective reaction to the colonizing aspects of outsiders coming in to effect policies to their profits (118), and thus, a singular rhythm lies in the power of farmers finding a way to farm their land.

In the fishing commercial, the son's Ulyssean return home seems to co-opt pathos-inducing Wendell Berry-esque images of the home place. The story also points to the complexity of the return home, of a healthy intergenerational relationship with the region, a perfect fit to ensure that our waters will remain clean. However, an awareness of exigency and reverence helps to reveal what lies hidden in the placid water. Not only can Pennsylvania be considered a veritable energy colony, the irony is that the gorgeous placid water that father and son were casting into was *already* polluted with mercury, which comes from numerous coal-fired plant sold to the local citizens due to our exigency for jobs. In Pennsylvania, a fishing license comes with a warning that it may only be safe to eat one fish a week due to high levels of mercury, PCB's and other contaminants ("Fish Advisory"). With this knowledge, the pristine seeming water where the father and son fish becomes murky as hidden agendas are revealed in the holistic images of this econ. In this area the vision has been led by industry-created false exigencies of relatively short-term jobs.

However, beyond the illumination of corporation and political collusion, this is also a call for all of us who live in this region to recognize our complicity within the tradition of colonization. Gregory Ulmer helps to understand the meaning of this complicity. In *Heuretics*, Ulmer provides a method of assimilating theory through invention rather than interpretation (3) by inventing alternative scenarios that tend to draw out the shadow realm of a particular scenario. Drawing from Derrida, Ulmer re-envisions the forgotten concept of *chora* separating it from *topoi*. *Chora*, the place between being and becoming, particularly suits this discussion as it deals with the subjective realm of how place interrelates with emotion; in other words, it can work to explain sacred space.

Thus, the invention here, the singular rhythm, requires the notion of *chora* and the need to take the responsibility to understand our complicity with the energy extraction industry. This recognition not only circumvents industry deflection of blame but circumvents a common question asked by industry representatives, "where do you think your electricity comes from?" This question is often used to make concerned citizens appear naïve. As a culture, we need to be able to take control of this question by understanding that in this realm of *chora*, the alternative scenario is that the politicians, the companies, the father and son, and all of us in this region float on the same beautiful, yet ruined, poisoned lake. We need to recognize that the 1,300 plus Pennsylvanians who die every year due to diseases from the burning of coal ("Death and Disease") float with us. They float with us because they were sacrificed in an abject way to enable us to continue our use of electricity.[3] The awareness created with the concept of eco-see can act as synaesthetic spectacles allowing us to witness underlying horrors. Once this awareness is created, the exigency then moves beyond the taking advantage of the economic opportunity of a lifetime, or the idea of drilling or not drilling; the exigency moves to the question of how can we change this nightmarish legacy, this stygian waterway upon which we all float? Further, what happens we might ask when the "godsend" that is gas drilling goes wrong? Can we drill our way away out of the coal era, as suggested by current government officials? If drilling can be done safely, why the Psy-Ops? Why the duplicity? Why the irreverence of elected officials and industry representatives?

Out of these questions, perhaps a more genuine dialogue can spring from true reverence and *chora*, one that promotes a healthier long-term rhetorical vision of place. A son returning home to work on a renewable energy project, helping local farms become holistically and energy independent by turning themselves into mini power stations, while setting up sustainable systems, such as food cooperatives and farmer's markets; an Oil and Gas act that protects all citizens, not just corporate citizens and leaseholders. This is not a vision that denies all drilling but one that perhaps truly understands drilling as a bridge to renewables, and perhaps this begins with the retrofitting of local coal-fired plants to burn natural gas which would keep the both the gas and jobs here. This is a beginning, a blossoming. This blossoming—a singular rhythm that involves the recognition of complicity; any movement or escape from the tradition of energy colonization—must occur at the local level, through dialogue, not separate monologues. Understanding reverence and exigency can help localities extend beyond the tradition of colonization to reach full potential.

NOTES

1. I primarily write from the context of western Pennsylvania, a region comprising twelve counties that surround Pittsburgh, that is my home place and the primary target audience for the range resource ads. However, most of what I write could be applied to any rural region of Pennsylvania and Appalachia. For instance, some of the biggest problems with fracking have occurred in Dimmock, Pennsylvania which is in the north-central region of the state
2. *Gasland* is the 2010 documentary that involved director Josh Fox traveling across the country interviewing residents who suffered numerous health and other problems after their land was fracked. In terms of shock value, the film depicted numerous residents lighting their contaminated water on fire.
3. This idea of abject sacrifice is inspired by Ulmer's idea of an Electronic Memorial discussed in Morey's and Guest-Jelly's 2011 4C's presentation as cited in the following Works Cited list.

WORKS CITED

Abbey, Edward. *Appalachian Wilderness: The Great Smokey Mountains*, Photographs by Eliot Porter. New York: Dutton, 1970. Print.

Berry, Wendell. *The Long-Legged House*. Washington: Shoemaker, 2004. Print.

Bitzer, Lloyd. "The Rhetorical Situation." *Philosophy and Rhetoric* 1:1 (1968): 1–14. Print.

Bryzcki, Elaine. "Explore Shale: An Exploration of Natural Gas Drilling and Development in the Marcellus Shale." Public Media for Understanding. Penn State University. 3 Jan. 2012. Web. 13 Feb. 2012.

Cawley, James. Indiana County Marcellus Shale Task Force. Kovalchic Center, Indiana, PA. 26 May 2011. Keynote Speech.

"Death and Disease from Power Plants." *Clean Air Task Force*, n.d. Web. 28 Dec. 2011.

Di Paola, Mike. "Fracking's Toll on Pets, Livestock, Chills Farmers: Commentary." *Bloomberg*, 8 Feb. 2012. Web. 11 Feb. 2012.

Dobrin, Sidney I. and Sean Morey. Eds. *Ecosee: Image, Rhetoric, Nature*. Albany: State U of New York P, 2009. Print.

Doyle, Arthur Conan. "When the Earth Screamed," n.d. Web. 2 Jan. 2012.

"Exigency" *Oxford English Dictionary: The Definitive Record of the English Language*. Oxford: Oxford UP, 2012. Web. 2 January 2012.

Explore Shale: An Exploration of Natural Gas Drilling and Development in the Marcellus Shale. Penn State Public Broadcasting, 2011. Web. 4 Jan. 2011.

Foucault, Michel *History of Madness.* Ed. Jean Khalfa. Trans. Jonathan Murphy and Jean Khalfa. London: Routledge, 2006. Print.

Friedenberger, Amy. "How the Shale Bill will Change Zoning Control." *Pittsburgh Post Gazette.* 9 Feb. 2012. Web. 11 Feb. 2012.

Howarth, Robert W., Renee Santoro, and Anthony Ingraffea. "Methane and the Greenhouse-Gas Footprint of Natural Gas from Shale Formations." *Climatic Change.* (2011) 106: 679–680. Web. 30 Dec. 2011.

Javers, Eamon. "Oil Executive: Military Style 'Psy-Ops' Experience Applied." *CNBC*, 8 Nov. 2011. Web. 2 Jan. 2012.

Kelsey, Timothy W., et al. "The Economic Impacts of Marcellus Shale in Pennsylvania: Employment and Income in 2009." *Marcellus Shale Training and Education Center.* Penn State College of Technology and Penn State Extension, Aug. 2011. Web. 30 Dec. 2012.

Killingsworth, Jimmie M. and Jacqueline S. Palmer. "Afterward." *Ecosee: Image, Rhetoric, Nature.* Ed. Sidney I. Dobrin and Sean Morey. Albany: State U of New York P, 2009. 299–309. Print.

Lewis, Helen. "Introduction: The Colony of Appalachia." *Colonialism in Modern American: The Appalachian Case.* Ed. Helen Matthews Lewis, Linda Johnson and Donald Askins. Appalachian Consortium: Boone, NC, 1978. Web. 13 Feb. 2012.

Marcellus Money. Common Cause Pennsylvania. 31 Dec. 2011. Web. 13 Feb. 2012.

McGinnis, James. "Cawley: No Evidence of Pollution from Fracking." *PhillyBurbs. com.* 5 June 2011. Web. 12 Oct. 2011.

Morey, Sean. "A Rhetorical Look at Ecosee." *Ecosee: Image, Rhetoric, Nature* Ed. Sidney I. Dobrin and Sean Morey. Albany: State U of New York P, 2009. 23–53. Print.

Morey, Sean, and Nic Guest-Jelly. "Relating the Disaster: Mapping the Spill, Mapping Ourselves (a MEmorial)." Conference on College Composition and Communication. Atlanta Marriott Marquis, Atlanta, GA. 7 April 2011. Conference Presentation.

Muckelbauer, John. *The Future of Invention: Rhetoric, Postmodernism, and the Problem of Change.* Albany: State U of New York P, 2008. Print.

MyRangeResources.com, n.d. Web. 20 May 2011.

"My Homecoming." Video. *MyRangeResources.com*, n.d. Web. 20 May, 2011.

"My Opportunity." Video. *MyRangeResources.com*, n.d. Web 20 May 2011.

O'Dell, Tawni. *Coal Run.* New York: Viking, 2003. Print.

Osborn, Stephen G., et al. "Methane Contamination of Drinking Water Accompanying Gas-Well Drilling and Hydraulic Fracturing." *PNAS* 108:20, 17 May 2011. Web. 13 Dec. 2011.

"Pennsylvania—Rich in Resources." *Keystone Energy Forum: Industry Insider.* Keystone Energy Forum, n.d. Web. 3 Jan. 2012.

Perry, Simona. "'It's Like We're Losing Our Love': Documenting and Evaluating Social Change During the Marcellus Shale Boom from 2009–2011." *Center for Instructional Development & Distance Education.* U of Pittsburgh, n.d. Web. 8 April 2012.

Phillips, Susan. "Could Fracking Earthquakes Shake Pennsylvania?" *State Impact.* National Public Radio, 23 Jan. 2012. Web. 25 Jan. 2012.

"Safety First, Fracking Second: Drilling for Natural Gas has Gotten Ahead of the-Science Needed to Prove it Safe." *Scientific American.* 19 Oct. 2011. Web. 13 Feb. 2012.

Schell, Eileen E. "The Rhetorics of the Farm Crisis: Toward Alternative Agrarian Literacies in a Globalized World." *Rural Literacies.* Ed. Kim Donehower,

Charlotte Hogg, and Eileen Schell. Carbondale: Southern Illinois UP, 2007. 77–119. Print.

Silver, Jonathon. "The Marcellus Boom Origins the Story of a Professor a Gas Driller and Wall Street." *The Pittsburgh Post-Gazette*, 3 March 2011. Web. 3 Feb. 2012.

Southwell, Steve. "Fart Fairies Responsible for Contamination, Say Drillers, Railroad Commission." *The Lewisville Texan Journal*, 12 Dec. 2010. Web. 3 Jan. 2012.

Steingraber, Sandra. "The Sound of Migration." *Orion* (Autumn 2001). Web. 13 Dec. 2012.

Stranahan, Susan. "Marcellus Rush Echoes History of Recklessness." Editorial. *Philly Burbs.com*, 20 Jan. 2012. Web. 9 April 2012.

Ulmer, Gregory L. *Heuretics: The Logic of Invention*. Baltimore: Johns Hopkins UP, 1994. Print.

Walls, David S., and Dwight B. Billings. "The Sociology of Southern Appalachia." Sonoma State, n.d. Web. 2 January 2012.

White, Jeremy, et al. "Toxic Contamination from Gas Wells." Map. *New York Times*. 26 Feb. 2011. Web. 10 Oct. 2011.

White, Mary Jo. "Marcellus Misinformation: Natural-Gas Drilling is Tightly Regulated in Pennsylvania." Editorial. *Pittsburgh–Post Gazette*, 23 June 2010. Web. 30 December 2011.

Woodruff, Paul. *Reverence: Renewing a Forgotten Virtue*. Oxford: Oxford UP, 2001. Print.

3 Sense of Place, Identity, and Cultural Continuity in an Arizona Community

Deborah L. Williams and Elizabeth A. Brandt

The predominantly Latino town of Superior, Arizona is located approximately an hour from metropolitan Phoenix, in east central Arizona. A former copper mining town, it has endured multiple cycles of boom and bust. The Magma Mine closed in 1995 leaving behind an economically and socially devastated community. This small, declining community is now the site of controversy over a proposed new copper mine to the east of the town. Resolution Copper Company (RCC), formed by Rio Tinto and BHP Billiton, seeks to extract copper from one of the largest deposits in the world, some 7,000 feet below the surface of pivotal landforms for the community, Apache Leap, and the conservation area of Oak Flat. Situated both on the mountainous edge of the Sonoran desert and on the cusp of new mineral development and technology, Superior is a place of competing identities, histories, and diverse meanings. The new mine would significantly alter the physical and cultural landscape of the region through the infusion of new jobs and funds into the community but at the expense of scenic and historic places of immense significance to some stakeholders. New mining will require that the Federal conservation area (Oak Flat) be moved to the private ownership of the mining company through a land exchange for other environmentally sensitive lands within Arizona. This requires congressional action to proceed and has been blocked in Congress for years, although the bill passed the House in 2011 for the first time.

The Superior landscape holds meaning for a variety of persons: Latino and non-Latino peoples, Yavapai people, Western Apache people, rock and boulder climbers, wildlife enthusiasts, miners, artists, and multi-national mining corporations. Mining brought Mexican miners, and the town has retained its Latino majority, roots, and culture. Other ethnic groups did migrate to the community but have remained a minority presence in the town. This cultural landscape is embedded in a stunningly beautiful physical landscape. It is unique in its landforms, flora, and fauna, and this impacts both the physical and social structure of the town. The town is hemmed in by U.S. Forest Service land, which limits expansion, but this same landscape provides a diverse set of recreation opportunities. Thus the cultural and physical landscape both constrains growth and expands activity. Despite a population decline of approximately forty percent in the last two decades, the community has received a small influx of recreationists and artists to experience the rugged beauty and active lifestyle of the community. Superior is thus home to a multi-ethnic community who experience diverse senses of place. These understandings, which arise through personal interaction within places, are embedded in the past and present, intensely local yet connected globally, both culturally and physically. The immigration of people and large national and international companies into Superior has brought new ideas, customs, and established links across communities and nations. Historically, the indigenous residents of the area, Yavapai and Apache, were displaced from the area in the nineteenth century, limiting access and interrupting cultural ties to their ancestral homelands.

As a place of both physical and rhetorical struggle, Superior embodies many of the tensions in our modern world. An interdisciplinary study was undertaken by a team of socio-cultural anthropologists and an ethno-geologist to understand how different senses of place, identities, and personal histories intersect when faced with new development that is massive in physical scope yet limited in financial viability to approximately forty-plus years.[1] Although many peoples hold cultural and historic ties to the area, this discussion focuses primarily on the Latino and Anglo inhabitants who form the recent past and present. The salient question for the stakeholders within Superior is: can a sustainable future be crafted which supports sense of place, identity, and cultural/community preservation? Will this future be inclusive or will some community members be marginalized or excluded?

The authors posit that individual and communal senses of place, along with the meanings and attachments held by those persons, influence how the resumption of mining is perceived by the Superior residents. We assert that those persons, who strongly identify with mining, and the mining history of the community, will be favorably disposed to mining resumption despite its limited economic time frame. Given that the 1995 mine closure threatened the continued existence of the town, the new mine is seen as potentially saving kin, place, and community. Conversely those individuals who migrated to the town for different reasons, or who are only lightly integrated into the community will express opposition to the new venture. We would expect that cultural ties, memory, and attachment would play a role in this determination.

This chapter continues with a discussion on the nature of place. How do we understand place, and how are place meanings and attachments constructed through our interactions within places? We then proceed to the controversy within Superior, a local issue reset within the national rhetorical frame of job building and economic recovery. The community of Superior is then presented, along with a discussion on residents' place attachments and meanings and how these entailments relate to opinions on the mine.

PLACES

We live, work, learn, play, love, and dream in places. In a contemporary world, our connections to places often seem tenuous or, as Semken and Brandt note, we must create new ties. Modern life is fast-paced, mobile, and information packed, yet place remains a cornerstone of human life. Just as we are all different, there are different types of places: places of memory, community, gathering, homecoming, healing, despair, war, death, power, work, solitude, and seeking. These places are interwoven into the fabric of our social lives and indeed are integral to its possibility. Our social activities are always placed, whether we overtly acknowledge them or not. Thus, places are situated spatially, temporally, and socially. *Sense of place* is a concept used to convey the interaction between people and their places and denotes the personal and emotional connections, which we as individuals and communities create and hold with places. This interaction creates place meanings and attachments that extend through time and space.

Places are not just sites but are dynamic processes, both physical and cultural. They are constructed and experienced by the peoples who live within them. Scholars such as Basso, Casey (*The Fate*), and Ingold advocate place as both socially constructed and materially experienced. Perception, activity, and dwelling are social experiences within specific places that result in the social and physical construction of world (Basso; Feld). Edward Casey asserts that we know, experience, and perceive in particular places, and thus knowledge of place is an ingredient in perception itself ("Body" 180). This dialectical experience of place invests it with meaning. Thus, places mediate our experience in the world helping us to organize reality (Casey "Body"). In this sense then, both persons and places emerge through activities within the landscape. Places are ultimately engagements; the inter-personal interactions with the specific environment in which we act (Bender 2).

This reciprocal experience creates and sustains both the place and the individual as noted by Basso (107–108) and Feld (99). Identity is enacted, embedded, and constituted through activities and conceptualizations held by individuals within specific places. This process is particularly apparent in the community of Superior, where social identities and senses of place are linked to dwelling and working within the landscape.

Places are also narratives. Discursively created and sustained, they are inherently dialogic. They come into being through discursive acts that are also social acts, whether solitary or communal (Rodman 642). Discourse, in and about our places, helps shape our experience, relationships, interactions, understandings, and knowledge. Places can be multi-vocal, multi-dimensional, multi-local, multi-cultural, and multi-historical (647). Places are individual and communal, local and global, self and other, historical, contemporary and future. In this sense places, especially contested places, are processes—verbs rather than nouns in their apprehension (multisensorial) and their understanding (polysemic). As different peoples hold different senses of place, conflicts may arise. Contested places are a result of humanity's lived experiences. Meaning making and identity are translocal processes. They are negotiated engagements and social interactions within, across, and about places.

Contested places reflect not only differences in meaning and use, but also differences in worldview and cultural commitments, normative and epistemological postulates, and power differentials. Histories inhabit landscapes (McNely 103). Power circulates within, through, and between places and their people. Sometimes places can structurally empower those who dwell within them, as in the town of Superior, where the mining infrastructure has created both physical and social structure. As Bender has previously asserted, contested places are concerned with relationships, power, and the politics of placement (10). The historical social positioning of Superior's Latino miners as second-class citizens via structural and discursive components within the town was an example of the legitimization of power and social order. This process of inclusion and exclusion cemented cultural cohesiveness and community within the Latino community, but kept them outside the wealthier, more politically powerful Anglo community who managed the Magma Mine. Some townspeople are wary of change and want life to stay the same, but with a restored community. For Superior, new development means that community identity is facing new challenges as it moves into the twenty-first century.

NEW DEVELOPMENT

Endres and Cook argue that "[t]he rhetorical deployment of place is a common tactic for social movements. They point out that protest groups use place attachments and memories in "place as protest" (258). They distinguish two types: "place-based arguments" and "place as rhetoric." A "place-based argument uses a description of a specific place as support for an argument" (264) and is often phrased as "an appeal to sense of place or saving a place" (265). "Place-as-rhetoric refers to the material (physical and embodied) aspects of a place having meaning and consequence, be it through bodies, signage, buildings, fences, flags and so on. Unlike place-based arguments that may invoke a non-present place to support and argument, place as rhetoric assumes that the place is rhetorical" (265). This is a useful perspective for examining some of the rhetoric surrounding this issue. We see both of these uses of place in this case.

Protests toward mine development were framed as losses of irreplaceable places by various groups of stakeholders—community members, climbers, native people, and conservation organizations—and an appeal by each group of protestors to save the places. These are all place-based arguments. News of the possibility of a new mine were initially just rumors. It appears that the climbers were the first group to confirm that a new mine might develop and contacted the Forest Service to verify this. The potential mine threatened areas such as Queen Creek Canyon, Oak Flat, and Devil's Canyon through loss of public access, loss of riparian areas, wildlife habitat, and water itself. The area has "over 2000 bouldering problems and established rope climbs." The Mountain Project website says that the Access Fund called the issue "the single largest loss of climbing in the history of the United States." The Fund represents over 2.3 million climbers and works to keep climbing areas accessible. Local climbers organized themselves online and began a campaign, which focused upon the irreplaceability of bouldering and climbing areas. After eight years of fighting and negotiations with RCC, the company has agreed to allow climbers to access their property through a permit system.

The San Carlos Apaches also protested citing their cultural ties to areas that would impact Devil's Canyon (the Apaches call this Gaan Canyon, which refers to the Mountains Spirits there—the English name is pejorative). They also expressed concern about Oak Flat and Apache Leap, which could subside as a result of underground mining and also discussed the irreplaceability of the area.

Conservation groups such as the Maricopa Audubon Society, the Center for Biological Diversity, the Sierra Club, and other groups used the rhetoric of conservation and environmentalism and the impacts upon very limited water resources in the area, and the provision of rare wildlife habitat. Other environmental groups also raised protests taking stances favoring continued protection of the Oak Flat area. When the land exchange bill came into Congress, its passage was hampered by federal charges placed against its sponsor, Representative Renzi, who came to trial for criminal corruption for attempting to extort land developers and copper mining companies for congressional influence. This issue is not yet resolved. All of the groups used place-based rhetoric.

Many of the groups also used place-as-rhetoric. One group had a climber place a banner high on the side of a cliff in Queen Creek Canyon saying, "Save Oak Flat." The Apache people held meetings at Oak Flat and in 2012 they had the Forest Service temporarily close the area in order to have a four-day Sunrise Dance or puberty ceremony for a girl coming of age. Coalitions of environmental groups, the Concerned Citizens and Retired Miners Coalition, and the Arizona Mining Reform Coalition joined together to hold a potluck in Oak Flat. These and many more activities in place allowed both types of place rhetoric to build.

In contrast to the protest groups, the mining company's rhetoric was focused on the jobs the project will bring to Arizona. This bill was touted as a real "jobs" bill at a time jobs were desperately needed, but the number of jobs has fluctuated significantly over the years. The recent downturn in the economy propelled the controversy onto the state and national stage as RCC and state officials began reframing the project within the national debate on job construction and economic re-structuring and recovery. A glossy Project Profile produced by the company early in the pre-feasibility phase projected more than 1,000 construction jobs and a minimum of 400 full-time positions.[2] During the most recent hearings on the new land exchange bill, H.R. 1904, this number increased to "3,700 badly needed mining jobs," according to RCC's Vice President Jon Cherry (2011).

As of the end of 2011, RCC has said that there are currently 100 RCC employees and 400 contractors working. In addition, because of the depth of the mine at 7,000 feet and the extreme heat and poor conditions for workers, most of the mine operations will be conducted by mechatronic automation, including ore trains, thus reducing the need for human jobs in the mine. However, Cherry's testimony focused on the environmental logic of the land exchange that will exchange environmentally sensitive parcels within Arizona for Tonto National Forest land, on the economic impacts, upon jobs, and concluded with responsible mining on lands within an area that has been mined for over 100 years. Cherry's argument was powerfully constructed and resonates with issues important to legislators and to the public at this point in time. The testimony of Jon Cherry clearly aligns perfectly with the national rhetoric of jobs, which dominated much of the public political discourse in 2011. Locally, the argument positions the project with historic ties to mining and with Hispanic mining identities. This bolstered support for the mine because it correlated with Latino senses of place and cultural preservation.

SUPERIOR, PAST AND PRESENT

Whereas the reframing of the project strongly resonated with many in Superior, it was the cultural and historical ties to mining that were rhetorically displayed. As we have noted, dwelling in a place is at once mindful and subliminal, rooted in history, memory, and daily acts of living that embed us within our places. Identity is thus intertwined with places. In the town of Superior, history plays an enormous role in both the physical structure and emotional character of the town. Reasons for dwelling in Superior are linked in substantial ways to opinions on mining and senses of place. These reasons center on employment, historical links, familial connections, and scenic beauty. There is a considerable amount of overlap, and hence tensions, as these understandings collide with development plans.

Superior is a mining town; mineral ore is ground into its bones. Mining has been a source of identity, often expressed in the phrase, *somos mineros* ("we are miners"). Historically Mexican laborers came singly and in family groups over 100 years ago to work the mines and remained to raise their families throughout generations. They, along with a few Eastern European and Mediterranean immigrants, established a vibrant community that existed alongside the Anglo management. Although the Latino community in particular suffered social, wage, and physical discrimination, the community nonetheless provided the community with relative prosperity, community, and connection. Mining, a translocal and global process bringing in outside management by global mining giants, cultural norms, and economic forces, nonetheless has resulted in a highly localized and enclosed community, deeply attached to the town and its surrounding landscape.

Memories of this time period are ever present in the physical structure and layout of the town and in family histories and personal narratives. This historical connection to mining is linked to memories of a vital community and town, which translates in the present to opinions on commercial development and the mine. Mining is still viewed as a respected and valued occupation, one that provides a good income and enables families to educate their children. This mining history is celebrated in the yearly Apache Leap Festival, which has mining competitions. Although Superior has long been noted as the town that has produced the most highly educated, successful, and prominent Latinos in Arizona, this has often led to their exodus from the town for employment. Strong family ties remain with some occupational migrants returning after successful careers to become involved in the community through local politics and civic development. The Superior Chamber of Commerce unifies the Latino and Anglo communities in working for increased development and jobs. The town has also recently created an economic development corporation to further this agenda and decrease the reliance on a single industry if the mine does go through.

PLACE MEANINGS AND ATTACHMENT

Superior is a nexus, a place of complex and multi-layered identities. Latino families, artist communities, unemployed poor, mining management, retired miners, disaffected youth, newcomers/outsiders, and Superiorites (long-term residents) all share in Superior-as-place. Family is both a strong inclusive element and a divisive one. Most of the long-term residents are Latino, second- and third-generation residents who are invested in the community. The town's cultural activities, although place-centric, often revolve around Latino holidays and activities. These familial and ethnic connections bind much of the community together despite loss of employment, residents, and infrastructure. The town has a significant drug problem and drug-fueled burglary, primarily with the young and unemployed, which have devastated the town socially and economically. Residents complain that crime is left unchecked due to familial ties, which extend into the judicial and legal structures. Despite the lure of employment elsewhere, many young community members remain to engage in specific meaningful places within the town and the surrounding landscape.

Many of these places are directly related to mining activity. The town is surrounded by a minescape of tailing piles, a smelter chimney, old buildings, and old mining scars on the surrounding mountains. When RCC took control of the old Magma Corporation property, it planned to raze the old buildings and tower, but the community chose to retain them, especially the smelter chimney, as a reminder of the mining legacy and town identity. "The Hollies," the mounds of mine tailings to the north of town, was a popular playground for the young people of the community. These individuals tell stories of returning home with orange-stained legs from long days spent on the mounds. Water-filled abandoned shafts and caves were frequented on hot summer days. Clandestine parties were arranged in "The Jungle," a wooded area near the Queen Creek. Town members relate memories of family gatherings at the now endangered Oak Flat area. *Pascua* ("Easter season") is the most important time in Catholic religious practice. Hispanic families often camped for days and celebrated Easter at Oak Flats. Some residents walk or bike to the area daily to experience the beauty of the area. These places form part of complex identities and memories, tied to mining, Hispanicity, and rural community life which forms the core of Superior-as-community identity.

The physical landscape thus contributes to community identity and sense of place. Long-term residents can map their histories upon the landscape and often lay claim to nearby landforms. Apache Leap is a source of pride and identity for the community. Residents consistently recite the "Apache Tears" tale in which Apache warriors fell to their death, driven to the edge by the U.S. military. The Apache women's "tears," obsidian pebbles, are found throughout the area.[3] Oak Flat is a place of gathering, for international recreationists, and for Yavapai and Western Apache communities both past and present. Landforms, monuments, and meaningful places within the landscape are repeatedly pointed to as examples of the beauty and singularity of Superior. Newer residents consistently cite the scenic vistas, mountains, and desert as the primary reasons for migrating to Superior. Thus community and landscape combine to create a lifestyle that is a strong motivating factor in continued residence. A small town attitude with perceived safety, conviviality, and community ties are the most cited advantages of town life. Outdoor activities and a pace of life that appreciates scenic beauty and direct environmental experience are sources of pride for town residents. The lack of amenities, vital services, and jobs is often tempered by a focus on the advantages of living in "the most beautiful place" in Arizona.

OPINIONS ON THE MINE

This emphasis on landscape as meaningful and valued complicates attitudes toward the resumption of mining. A majority of the residents approve of the new venture, especially the close-knit Latino community. Some speak of the proposed new mine with a sense of ownership, "our mine," and resent outside opposition. Mining and the anticipated economic revital-ization it will provide are seen as instruments in maintaining family con-nections and physical community structure in the face of ever-increasing exodus of residents seeking jobs. However, an important and historically resonant portion of this landscape will become inaccessible or lost with the new mining operation. Opinions on new mine development demonstrate some degree of coherence along the dimensions of ethnicity and length-of-residence. On the one hand, both long-term and new residents wish for an economically vital and sustained town; on the other, new residents are more likely to oppose the new mine as being detrimental to the scenic and environmental values of the area. Many express concerns about pollution and the increased traffic of heavy trucks.

Long-term Latino community members support new mining citing eco-nomic vitality, social stability, and *minero* identity as valid reasons for its resumption. Although there is concern for health and the loss of important places, these individuals feel that without new mineral development the town will die. Loss is therefore preferable to community death. Revital-ization of the downtown district is particularly anticipated by this group. Latino residents speak of the days when bars, shops, restaurants, and even bordellos abounded creating a bustling district and gathering place. Vivid memories of events and people remain placed within this now crumbling and often derelict area. This attachment to specific places and to Superior as a whole is expected in long-term residents and was consistently dem-onstrated with the high place attachment and meaning scores collected from Latino inhabitants. Long-term Anglo residents present a more mixed response. New economic development is viewed as necessary, but a return to a "one-employer" town is seen as an insecure base—hasty and ultimately negative. Some long-term Anglo residents criticize community members as frightened of change and wishing for a return to the mining past, of being "taken care of" by the mining company. Anglo residents, who most often are more prosperous property owners or retirees, want diversified economic revitalization and may exhibit only tenuous links (if any) to the past.

New residents are of two types primarily: artists and retirees. Both groups tend to be White Anglos, although Latino artists are present. Artists in general are opposed to the new venture and their reasons are tied into the scenic beauty and meaning of the landscape. They cite environmental degradation, pollution, and health issues (all previously experienced in the town) as reasons to delay or oppose the new mine. These new residents feel little connection to the mining past. Superior for this group is the landscape and viewscape, coupled with small town life. Retirees are somewhat mixed in their responses. Many have chosen to reside in Superior for economic reasons coupled with the scenic beauty of the landscape. The dearth of amenities and high crime rate are seen as significant problems that might be alleviated with the new mine. These newer residents also scored high on place attachment and meaning surveys. Perhaps the main difference between newer and older inhabitants is that their lived experiences differ and so therefore does the nature of their attachments and meanings. Community, family, and history are embedded in places for the Latino members of the community whereas new residents emphasize more individual meanings and attachments.

Differences are manifested in the cultural landscape that interpenetrates individual and group identities, although the scenic beauty and physical landscape of the area binds these disparate groups together. These differences, encompassing different senses of place, are expressed through rhetorical struggle in the current controversy. Cultural context and identity are integral to these senses of place. For many long-term Latino residents and some Anglo residents, Superior is a mining community (rather than a mere town). Mining supplies employment builds and maintains infrastructure, and facilitates (indeed even penetrates) family and community interactions within this context. Other residents, often recent or those inhabitants with less positive views on mining, experience Superior as a disadvantaged town set in an incredible landscape. Mining for these individuals can only harm Superior-as-place, particularly in the long term.

CONCLUSION

The future of Superior is complicated by diverse histories, identities, and meanings. Mining founded and fostered the community but was responsible for its devastation, both physically and economically. It was both a cohesive force and a divisive instrument among the inhabitants, sustaining the town but separating communities within it. This complex history is part of the social fabric of the community and a component of individual and community identities. The resumption of mining in Superior continues this pattern of cohesion and division. Place is both a site of protest and a reason to protest, hence the use of both types of place rhetoric.

Long-term Latino residents and intergenerational *mineros* see the new mine as a means to restore and maintain community viability and cultural continuity. Personal and family narratives, embedded in mining places and structures within the town, might be reclaimed and preserved by funds injected into Superior by RCC. Jobs within the community will reenergize and cement community ties, encourage residence, and revitalize physical and social structures.

Other long-term residents feel differently. Some miners and long-term residents strongly oppose the mine and the physical impacts upon the town that mining has created. Having lived through the cycles of boom and bust, coupled with environmental degradation, these individuals strongly identify with the more negative aspects of mineral development. Mining, although a valid livelihood, is a single economic driver, which will ultimately negatively impact community continuity.

More recent residents lack this "mining identity" yet strongly identify with the town and the surrounding area. These individuals often desire diversified commercial ventures in tourism, the Arts, and perhaps mining as the key to community viability. In a town often characterized as a "ghost town" new economic development is vitally needed.

The scenic beauty and recreational value of the landscape impact senses of place for all inhabitants. Dwelling in Superior is more than living in a mining town. Town residents consistently identify Superior as a richly meaningful and personal place. This encompasses the town itself, the surrounding desert and mountains, and the scenic vistas. New mining offers hope, but threatens current place attachments and meanings. Although many residents applaud the passage of H.R. 1904, the rhetoric of Superior residents centered more on a sustainable future than on jobs alone.

Superior is multi-vocal, multi-cultural, and polysemic, traversing the vibrant past and the impoverished present. The town and community exemplify the relationships that coevolved with and within places through time. This history embedded within places, community, and individuals, shapes identity, community viability, and meanings in the present as the town confronts the resumption of copper mining.

NOTES

1. This research was generously supported by a one-year fellowship in 2007–2008 from the Institute of Humanities Research to Elizabeth A. Brandt, School of Human Evolution and Social Change and Steven Semken, School of Earth and Space Sciences at Arizona State University. Deborah L. Williams served as the research assistant for this project. Research methods included ethnography, participant observation, interviewing, and surveys on place meaning and place attachment.
2. The number of years during which the project will actually mine has varied over the years from approximately forty early in the proposed project to the most recent figure of sixty-six years which is mentioned in the testimony of Jon Cherry, Vice-President of Resolution Copper Company in 2011.
3. Archaeological evidence shows that there was extensive Apache and possibly Yavapai Indian use and occupation of the Apache Leap escarpment. Some historical evidence as well as archaeological evidence, including Civil War–era military shell casings and a rifle stock from the same era, suggests that there was some type of conflict in the area. However, the Apache Tears legend appears to have been promoted by a family who owned the obsidian quarries in the 1940s.

WORKS CITED

AccessFund. Save Oak Flat. "Action Alert." 10 June 2011. Web. 3 Sept. 2012.

Basso, Keith. *Wisdom Sits in Places*. Albuquerque: U of New Mexico P, 1996. Print.

Bender, Barbara, ed. *Landscape: Politics and Perspectives*. Oxford: Berg Publishers, 1993. Print.

Casey, Edward S. "Body, Self and Landscape." *Textures of Place: Exploring Humanist Geographies.* Ed. Paul C. Adams, Steven Hoelscher, and Karen E. Till. Minneapolis: U of Minnesota P, 2001. 403–425. Print.

———. *The Fate of Place: A Philosophical History.* Berkeley: U of California P, 1997. Print.

Cherry, John. "Testimony of Jon Cherry, Vice-President, Resolution Copper Company, before the U.S. House of Representatives Subcommittee on National Parks, Forests and Public Lands concerning H.R. 1904 Southeast Arizona Land Exchange and Conservation Act of 2011." June 14, 2011. Web. 5 Dec. 2011.

Endres, Danielle, and Samantha Senda-Cook. "Location Matters: The Rhetoric of Place in Protest." *Quarterly Journal of Speech* 97.3 (August 2011): 57–282. Print.

Feld, Steven. "Waterfalls of Song: An Acoustemology of Place Resounding in Bosavi, Papua New Guinea." *Senses of Place.* Ed. Steven Feld and Keith Basso. Santa Fe: School of American Research P, 1996, 91–136. Print.

Feld, Steven, and Keith Basso, eds. *Senses of Place.* Santa Fe: School of American Research P, 1996. Print.

Ingold, Tim. *The Perception of the Environment: Essays in Livelihood, Dwelling and Skill.* London: Routledge, 2000. Print.

McNely, Brian J. "La Frontera y El Chamizal: Liminality, Territoriality and Visual Discourse." *The Responsibilities of Rhetoric.* Ed Michelle Smith and Barbara Warnick. Long Grove, IL: Waveland P, 2009, 96–114. Print.

Resolution Copper Mining. *Project Profile.* N.d.

———. *Economic and Fiscal Impact Study.* April, 2008.

———. *Project Profile.* January, 2009.

Rickus, John. "Testimony of John Rickus. President, the Resolution Copper Mining (LLC) before the U.S. House Subcommittee on National Parks, Forests and Public Land on H.R. 3301, Southeast Arizona Land Exchange and Conservation Act of 2007." 1 Nov., 2007. Web. 5 Dec. 2011

Rodman, Margaret C. "Empowering Place: Multilocality and Multivocality." *American Anthropologist* 94 (1992): 640–656. Print.

Salisbury, David. "Testimony of David Salisbury, President, Resolution Copper Mining, LLC before the U.S. Senate Committee on Forests and Public Lands concerning S. 3157, Southeast Arizona Land Exchange and Conservation Act of 2008 . 9 July 2008. Web. 5 Dec. 2011.

Semken, Steve, and Elizabeth A. Brandt. "Implications of Sense of Place and Place-Based Education for Ecological Integrity and Cultural Sustainability in Contested Places." *Cultural Studies and Environmentalism: The Confluence of Ecojustice, Place-Based (Science) Education, and Indigenous Knowledge Systems.* Ed. Deborah J. Tippins, Michael P. Mueller, Michiel van Eijck, and Jenifer D. Adams. New York: Springer. 2010, 287–302. Print.

Wilder, Nick. "Mountain Project. Oak Flat Action Alert: Save Oak Flat-Queen Creek Climbing." 21 Nov. 2011. Web. 3 Sept. 2012.

4 Mt. Taylor, New Mexico

Efforts to Provide Resilience to a Sacred Mountain Socio-Ecological System

Sally Said

In his introduction to *Speaking for the Generations: Native Writers on Writing*, Simon Ortiz, Acoma Pueblo poet, writes,

> Speaking for the sake of the land and the people means speaking for the inextricable relationship and interconnection between them. Land and people are interdependent. In fact they are one and the same essential matter of Existence. They cannot be separated and delineated into singular entities . . . [W]ithout the land there is no life, and without a responsible social and cultural outlook by humans, no life-sustaining land is possible. (xii)

The land currently in question is located on and near Mt. Taylor, an 11,301-foot stratovolcano in northwestern New Mexico, formed between two and four million years ago and now surrounded by mesas and lava flow. Among those who hold the mountain sacred, the people who speak for the sake of that land are the Pueblos of Acoma, Laguna, and Zuni; the Hopi Tribe; and the Navajo Nation. These five tribes have collaborated to nominate the mountain and surrounding area as a Traditional Cultural Property (TCP) with the New Mexico Department of Historic Preservation's Cultural Property Review Committee (CPRC). Such a designation would offer tribal leaders the right to be consulted in the state permitting process for major construction or mining within the area. The five tribes made the request for temporary listing on the State Register as an emergency measure to slow down the rush by uranium companies to gain fast-track permits for exploratory drilling. The CPRC approved temporary listing in June 2008, and permanent listing in June 2009. There was opposition to the two decisions from other stakeholders: uranium companies and those who owned land or had mineral interests in or near the proposed TCP, as well as the city of Grants and Cibola County, which stood to benefit economically from resumed uranium mining. Despite CPRC assurances that the TCP designation would not interfere with existing private property rights, other stakeholders, led by uranium companies, filed suit against the CPRC and Acoma Pueblo in district court, and in February 2011 the TCP designation was set aside.

In order to understand attempts to determine the future of Mt. Taylor and of those who depend on it to meet spiritual and economic needs, this chapter will rely on socio-ecological resilience theory from works such as the 2003 collection of essays edited by Berkes, Colding, and Folke, *Navigating Social-Ecological Systems: Building Resilience for Complexity and Change*, and on theories of social resilience set out by Kirmayer and colleagues in their 2009 article "Communicative Resilience: Models, Metaphors and Measures" and by authors in Goldstein's 2012 edited volume *Collaborative Resilience: Moving Through Crisis to Opportunity*, among others. It is the perspective of this chapter that social resilience is essential to ecological resilience in a system which links them, and that worldviews and shared knowledge as communicated through rhetoric can shape the social resilience, and thus the ecological resilience, of the Mt. Taylor socio-ecological system. Failure to share such knowledge and build community can also diminish social resilience and make the socio-ecological system more vulnerable to economic exploitation with negative social and environmental impacts.

Each of the five tribes identifies Mt. Taylor as part of its origin or migration story, site of cultural practice, home to sacred beings, and place of pilgrimage. Long before the Spanish called it *Cebolleta* ("wild onion"), or Americans named the mountain for President Zachary Taylor, it was known as *Tsoodził* ("mountain tongue") and *Dootł'izhii Dziil* ("turquoise mountain") to the Navajo (Linford 235), as *Kaweshtima* ("place of snow") to the Acoma, as *Tsiipiya* ("glancing touch") to the Hopi, as *Tsibina* ("forested mountain") or *Tse-pi'na* ("woman veiled in clouds") to the Laguna, and as *Dewankwi Kyabachu Yalanne* ("in the east snow-capped mountain") to the Zuni (Benedict and Hudson 18–23). For these five tribes, the mountain also provides an ecological niche for the growing of plants needed in healing and ceremonies. Furthermore, it is the principal source of water for the Acoma and Laguna Pueblos. Additional uses of the mountain are recreation, ranching, logging, and mining. Landowners/managers include the Bureau of Land Management, the National Forest Service, individual ranchers, and a Spanish Land Grant community. The land also contains among the richest uranium deposits in the United States, the Grants Mineral Belt. With rising demand, uranium companies have recently proposed the resumption of mining, discontinued in the 1980s when the price of uranium dropped. According to the Laguna Acoma Coalition for a Safe Environment, prior mining left un-reclaimed and abandoned mines, piles of mill tailings, and leach piles—sources of contamination of the soil and water. The coalition asserts that renewed uranium mining, a water-intensive process, will result in depletion and contamination of aquifers, surface springs, streams, and seeps.

Through the of lens socio-ecological resilience theory, linked ecosystems and societies, seen jointly as complex socio-ecological systems (SESs), undergo cycles of growth and collapse, followed by reorganization and renewal, in four stages: exploitation, conservation, release, and reorganization (Berkes, Colding, and Folke 17). This "adaptive renewal cycle" yields opportunities for intervention to guide the process of renewal, and to strengthen the mature system to provide greater adaptive capacity, defined by Walker and Salt in *Resilience Thinking* as the ability "to absorb disturbance and still retain its basic function and structure" (xiii). The ability to learn from disturbance and adapt, not to resist change and return to a prior equilibrium state, is the central characteristic of system resilience. The social and the ecological parts of the system are linked by long-term interaction. According to Walker and Salt, "the biophysical system constrains and shapes people and their communities, just as people shape the biophysical system" (34).

Systems that exhibit resilience have certain characteristics that not only provide for response to sudden disruptive events but also keep them on track in a constantly changing world. Increased intentionality and collaboration of human members enhance the adaptive capacity of an SES because complex issues can then "be dealt with by a network of loosely connected stakeholders located at different levels of society" (Walter and Salt 138). The resulting resilience may be considered an emergent property of a system, a benefit that cannot be predicted or understood simply by examining the system and its parts, but derives from its complexity and the internal interaction that takes place (5). Two types of memory also promote resilience. History, in terms of traditional ecological knowledge, recall of stories, and reenactment of ritual, as well as written records, helps to shape the path of the social system's adaptation to change. Similarly, an ecosystem's history of a place, its experience under conditions of change over time and the genetic resources of surviving species, can guide the ecological system in its response to current disruption. History, linked to place, is a record of learning, an embodiment of Keith Basso's finding with the Western Apache that "wisdom sits in places" (127). With access to this wisdom, those of the right sort of mind—a resilient mind—recognize the lessons encoded in places and use them for the benefit of the community in times of distress: "The social group survives, shaken, but whole, and the qualities of mind responsible for its continuation are made clear for all to see" (135).

The statements offered by the five tribes in their TCP nomination demonstrate a long period of interdependence and familiarity with the mountain and surrounding lands, a history in place over 1,000 years, in the case of Acoma Pueblo (Anschuetz; Ferguson; Hopi Tribe; Kelley, et al.; Zuni Tribal Council). Underlying the statements themselves are worldviews that include a tribal origin and progression through a series of worlds to a final emergence into the present one. For the Hopi and the Navajo, each world was destroyed or abandoned because of malfeasance by the ancestors (Hopi Tribe; Kelley et al.). The progression gave a sense of advancement, wisdom, and increasing responsibility to the chosen ones allowed to survive. In summarizing the Acoma tribal statement in the nomination document, archeologist Kurt Anschuetz notes the "sacred obligation that the Pueblo's people accept as stewards of this landscape in exchange for their inheritance" (10). This is to say, the Native cosmologies include an understanding of successive worlds, not unlike the adaptive cycle, with its development, destruction, and renewal as a regular part of life in the past, leading into the present world. The Creator (Hopi) or the Holy People (Navajo) exacted a promise from the inhabitants of each new world to respect Creation and the Creator, and to perform ritual acts and prayers to show gratitude. The prayers and rituals performed on Mt. Taylor are done to protect the Earth as the Native people know it, not just for themselves, but for all humans and for future generations. In the Laguna statement, several of the Laguna sources interviewed by Ferguson report that religious activities on Mt. Taylor are beneficial for the whole world (53).

Both diversity (with a variety of resources for adaptation) and minor disturbance (a force for change) are necessary to maintain flexibility that gives resilience. On the other hand, rigidity in the conservation portion of the adaptive cycle can actually limit learning and decrease adaptive capacity (Berkes, Colding, and Folke 23). For example, insistence on fire suppression and prevention has created fire-susceptible forests across the nation (Goldstein and Butler 339). Rigidity can also affect social resilience when a subpart holds to beliefs that make communication among members of the social system difficult and impede change. For some of the non-Native stakeholders, negative attitudes toward Native Americans stemming from the almost 500 years of struggle to appropriate indigenous land combine with a sense of property rights as natural and inherent and a worldview giving humans dominance over nature to produce rigidity. Reluctance to engage in public discourse with those who dominated them historically and the required secrecy concerning ritual matters also hinder a Native approach to Mt. Taylor issues that is more inclusive of other politically powerful stakeholders. According to resilience theory, destabilizing such an inequitable and harmful system might increase resilience in the long run, even if the transition requires temporary loss of system function (Goldstein 363). Having both factions of the social system sit to discuss its future would likely produce conflict, but might result in long-term gains in terms of trust and an ability to plan together. That no such community building took place in advance of discussions of TCP status made those discussions largely matters of defending positions and not discovering interests, whether common, differing, or conflicting. As Patton notes in his "Negotiation" chapter in *The Handbook of Dispute Resolution*, "the potential value inherent in shared or differing interest may be as large or larger than the value in dispute" (281).

The tribes oppose uranium mining on the mountain largely from experience. The shameful history of uranium companies in the region, active from the 1950s through the 1980s, that left in their wake contaminated land and water, radiation-related illness, and birth defects among former miners and their families. Several contaminated sites still remain in the Mt. Taylor area in addition to the un-reclaimed Mt. Taylor mine site, including the Jackpile mine tailings pile on Laguna Pueblo land (Laguna Acoma) and the L-Bar uranium mill tailings pile on the Cebolleta Land Grant on the eastern slopes of Mt. Taylor (Southwest Research Information Center 1). Mining on the Navajo Reservation left behind the same aftermath of un-reclaimed land, polluted water, radiation sickness, in addition to homes built unknowingly with radioactive materials, described in Judy Pasternak's exposé book *Yellow Dirt*, the title a translation of the Navajo name for uranium ore, *leetso*. In 2005, the Navajo Nation Council passed the Diné Natural Resources Protection Act citing prior "social, cultural, natural resource, and economic damage" from uranium mining and prohibiting extraction of dangerous substances that should be left in the ground, identified in "ceremonies and stories that have been passed down from generation to generation" (Navajo Nation Council 2).

In further recognition of the danger of uranium mining to sacred sites, on April 28, 2009 the National Trust for Historic Preservation named Mt. Taylor to its List of America's Eleven Most Endangered Historic Places:

> Much of the area is governed by the 1872 Mining Law, which permits mining regardless of its impact on cultural or natural resources, meaning that the U.S. Forest Service and other federal land management agencies lack the authority to deny mining applications, even if the application would adversely affect those resources." (Hays "National Trust" 1)

Listing Mt. Taylor as a TCP would require mining companies to obtain a standard permit and a full review by the state Historic Preservation Division, including consultation with the tribes, before exploratory drilling could begin (WISE). This is the best tribes can hope for until the laws regarding mining are changed. In her conference paper "Geographies of Mt. Taylor: The Legal, Political and Cultural Implications of Uranium Mining in One of New Mexico's Most Sacred Spaces," geographer Melinda Harm Benson notes that law affecting the case is overlapping and contradictory, especially as regards "the extent to which federal agencies can say 'no' to mining proposals, or at least place environmental management controls on mining activities" (8).

Because the CPRC nomination form does not deal with threats to the TCP, the nominating tribes can only describe the ways in which Mt. Taylor qualifies. There are Native advocacy groups, however, which deal more directly with the uranium issue, including the Multicultural Alliance for a Safe Environment (MASE), Eastern Navajo Diné Against Uranium Mining (ENDAUM), Sacred Alliances for Grassroots Equality (SAGE), the All Pueblo Indian Council, and the Laguna Acoma Coalition for a Safe Environment (LACSE). They have been joined by the Sierra Club; the New

Mexico Environmental Law Center; the Southwest Research and Information Center (SRIC), a research group that follows uranium issues; and the National Trust for Historic Preservation.

The difference in worldview between the TCP advocates and their opponents is evident in the statements by an organization called Citizens' Alliance for Responsible Energy (CARE), a pro-nuclear energy group, whose leader, Marita Noon, has also published self-help books from a conservative Christian perspective as Marita Littauer (Paskus 12). In response to permanent listing of the Mt. Taylor TCP, Noon, known for her hyperbole, writes that the decision was "New Mexico's 1 million acre land grab" (1). She continues,

> The sad truth is the decision is not in the public's best interest as the jobs and uranium would have been a major asset to both New Mexico and America. It can happen because the public wasn't interested . . . Most of us sat it out while the proponents pushed hard to get it through. The local citizens in Grants, the attorneys representing the locals and small mining companies, and a couple of interest groups (like CARE) were not enough to stop the wave of political correctness. (2)

Noon fears that the example of Mt. Taylor will encourage "mineral resource development opponents" to repeat the process elsewhere, because "those of us who value free-market principles, believe in private property rights, and support energy freedom" were sleeping (2).

In order to examine how the rhetoric used by members of the Mt. Taylor SES affects SES resilience, this work looks to forms of social resilience called collaborative, communicative, community, and consensus/dissensus, which all depend on rhetoric for their creation and operation, and which I term the four Cs. Goldstein's *Collaborative Resilience: Moving Through Crisis to Opportunity* reflects the fact that the contributors were among those invited to Virginia Tech a year after the 2007 murders to discuss "how collaboration in its many forms could promote resilience to crisis" (1). The types of collaboration (the first C) include "the stories of communities that have survived and thrived through adaptive consensus building and transformative social change, altering assumptions, behaviors, processes, and structures for the greater good" (2–3). Goldstein also develops the notion of communicative resilience (the second C), "a framework for communities to both define and pursue resilience through collaborative dialogue, rather than through expert analysis" (366). Communicative resilience results when collaboration "reshapes both the knowledge and the knowers" (368).

In developing a definition of the third C, community resilience, Kirmayer and colleagues conducted a 2009 study among the Aboriginal People of Canada, "Community Resilience: Models, Metaphors and Measures." Relying on resilience theory from the social and behavioral sciences, they note that the strengths of Aboriginal culture aid resilience at all levels of their society: individual, families, communities, and larger social systems (Kirmayer et al. 62). One strength is the emphasis on bonds with others and with place, "central to indigenous notions of identity and community" (65). Another is diversity, allowing for individual difference and the creation of alternative modes of response to outside pressures to assimilate. A dynamic property of systems, resilience can make use of diversity to insure the system's continuity "in ways that maintain its components but it may also transform or eliminate components" (72). Such systems need to establish contact horizontally with members of their own community ("bonding") and with other similar communities ("bridging"), and vertically with administrative structures that have control over their activities or projects ("linkage") (73).

The notion of achieving resilience through consensus (the fourth C) suggests that the end product of collaboration or communication be agreement on essential issues, but not to the point of rigidity. Given that a certain amount of the diversity of worldview and beliefs is desirable, disagreement is to be expected along the way. Dissensus, as defined by Ingham in "Landscape, Drama, and Dissensus: The Rhetorical Education of Red Lodge, Montana" is as important as consensus in assuring resilience (198). That is, it is important to know where disagreements lie, to allow all voices to be heard, and to work on developing action plans in areas where there is agreement. The interaction resulting from this work may help to resolve some of the remaining disagreement by building familiarity and trust, social networks, and good communicative practice, or "rhetorical health" (198). Community meetings at Red Lodge were facilitated by the Sonoran Institute of Tucson, Arizona, whose director, Luther Probst, believes that workshops based on a rhetoric of possible futures, better versions of citizens and the community, in advance of decision making, are much superior to a contentious public hearing process on proposed action (Ingham 206).

Any rhetoric of resilience, thus, would have to perform certain functions: create bonding (build communities, strengthen commitment of a small group to the project of joint planning), build bridges horizontally (allow for dialogue with other groups), and provide links to hierarchical governmental structures that determine policy. The community, rhetorically created, would have to share social knowledge, giving the possibility of consensus, discovering dissensus, and converting consensus to action. Efforts to limit harm to the Mt. Taylor SES began in June 2007, when the All Indian Pueblo Council passed a resolution calling for protection of Mt. Taylor sacred sites, in light of the drilling and exploration permitted by the New Mexico Energy, Minerals, and Natural Resources Department without consulting the nineteen pueblos and tribes affected (Manataka 2). After a lengthy collaboration to document tribal relationships to the mountain, the Mt. Taylor Ranger District of the Cibola National Forest declared Mt. Taylor eligible for listing on the National Register of Historic Places

February 4, 2008. The next step would be for a subset of the affected tribes to seek listing by the New Mexico CPRC on the State Register of TCPs. In so doing, they demonstrated bonding within tribes—bridging with other tribes, affirming their joint commitment to their shared sacred sites—and vertical linkage with a government agency, but failed to connect with other land and mineral rights owners in the region to discover common interests, ties to the land, and other bases of collaboration necessary to assuring social resilience of the SES and broader support for the TCP designation.

The criteria set out for eligibility for listing as a TCP on the National Register are also those used by the CPRC, and are provided in National Register Bulletin 15 and clarified in Bulletin 38: (a) properties associated with events that have made significant contributions to the broad patterns of the group's history (including oral history, National Register Bulletin 38: 11); (b) properties associated with the lives of persons significant in the group's past (including "gods and demigods who figure in the traditions of the group," National Register Bulletin 38:11); (c) architecturally significant man-made structures; and (d) properties that have yielded, or may be likely to yield, information to the group's history or prehistory. The Forest Service nominated the Mt. Taylor TCP under criteria (a), (b), and (d) above. Such an eligibility designation affords the TCP the same protections as an actual listing, under Section 106 of the National Historic preservation Act, because full descriptions of sacred sites required for actual listing could endanger them or violate tribal restrictions. The Forest Service nomination was prepared by archeologists Cynthia Benedict and Erin Hudson, in consultation with tribal representatives "regarding cultural values and uses of Mt. Taylor," beginning in the fall of 2007 (Benedict and Hudson 13). The Forest Service was very inclusive in trying to reach all indigenous groups with historical and cultural ties to Mt. Taylor. Of the sixteen tribes and thirteen chapters of the Navajo Nation contacted, a smaller group actually met with the authors: the Pueblos of Acoma, Laguna, Zuni, Jemez, and Isleta; the Hopi Tribe; the Jicarilla Apache Nation; and the Navajo Nation (14). The contents of the nomination document set the pattern to be followed by the tribes in their separate petition for temporary, and later permanent, listing as a New Mexico TCP.

Despite the tribes' success in collaborating to nominate the Mt. Taylor TCP for listing with the CPRC, with regard to the other stakeholders, several rhetorical missteps hardened opposition to the designation. If the only issue had been protection of sacred sites, then seeking TCP status would not have been controversial. It was clear that the sites met the TCP requirements for listing once the five tribes who petitioned the CPRC had documented their cultural and historical links to the sites. Given that there was another issue, uranium mining, someone—the tribes, the CPRC, a community agency—could have anticipated the need for consensus building in advance of hearings on the merits of the TCP. In a series of meetings, the divided community could have come together to discuss the potential impact of mining on the SES, to share cultural stories linking them to the land, making sure all voices were heard, and possibly discovering common interests before deciding on a plan of action. That is, the two groups could have joined to create social resilience and establish good communicative practice or "rhetorical health" (Ingham 198). However, no such meetings took place, and in the absence of wider consensus, the five tribes sought protection directly from the CPRC in February 2008, precipitating a crisis in an attempt to avoid one.

In a second misstep, the temporary listing was discussed not at a regular meeting but at an emergency meeting on February 22 (Manataka 3). Other stakeholders perceived the move to rush the decision as adversarial, and the issue became more polarized. The Cibola County Commission voted 4–1 in a very heated meeting in April 2008 to oppose designation, citing the negative economic effects of the requirement for "mining interests to obtain a standard permit and a full review by the state Historic Preservation Division before exploratory drilling can begin" (Manataka 2). Responding to a third misstep, to be repeated successively, on May 6, 2008, the New Mexico Attorney General's Office declared the temporary TCP listing invalid because the CPRC staff "had provided inadequate notice of the special meeting to property owners within the boundaries or near the TCP Property" (Manataka 2). With what the CPRC believed to be proper notification and distribution of an agenda, a meeting to grant temporary listing was again held June 14, 2008 and meetings for permanent listing May 15 and June 5, 2009. At the hearings, the CPRC received oral and written statements from nominators, government officials, and the public, then after deliberation, approved the listings.

The rhetorical malfunction of these meetings involved the casting of other stakeholders as outsiders, opponents of a proposal developed without their input. They could only respond to the designation proposal reactively, and had no chance to tell their own stories of their ancestors' hundreds of years on the land (in the case of Cebolleta Land Grant residents), to share their own knowledge of the ecosystem, to shape the proposal for protecting the mountain. In a failure to listen to the arguments of the opposition and hear the sense of injustice that underlay them, the CPRC assumed that the questions raised were matters of law, which could be answered by government experts (Final Order 9–10).

Despite legal clarifications, the opposing stakeholders did not accept the outcome of the June 5, 2009 CPRC decision, and sought ways to have it set aside by finding weaknesses with the decision itself and with the process. Still, no one sought to promote better communication between the two sides and to allow them to explore the desirability of uranium mining in the region, the underlying issue. On August 10, 2010, opponents of the nomination filed suit in the New Mexico Fifth Judicial District Court. The CPRC tried to remove the case to federal court, but the plaintiffs dismissed their federal claims assuring a hearing in state court. On the side of TCP supporters, a Brief of *Amici Curiae* was filed in September 2010 by the All Indian Pueblo Council, the American Anthropological Association, the Association on American Indian Affairs, the National Trust for Historic Preservation, the Sierra Club, and the Society for American Archeology (All Indian Pueblo Council et al. 2). Their intent was to reaffirm the final order, especially as it dealt with the issues of religion, culture, and history.

On February 4, 2011, Judge William Shoobridge found for the plaintiffs (Jaramillo 2). The judge heard the case in December 2010, brought by "private landowners on the mountain, several uranium mining companies, Public Lands Commissioner Patrick Lyons, and the Cebolleta Land Grant" (1). Petitioners' attorneys argued that the area was too large to qualify as a TCP and too large to be overseen by the CPRC. Due to clerical error, the final area was given as much larger than originally asked for. The judge suggested breaking the TCP into parts to be protected separately, finding that the TCP as a whole lacked integrity of location. He also found that again errors were made in notifying interested parties, specifically mineral owners, and that the Cebolleta Land Grant should not be included (Hays "Court Ruling" 1).

Failure to address the economic needs of the area as part of social resilience omitted from consideration documents such as a 2008 report prepared for the New Mexico Environmental Law Center, "An Economic Evaluation of a Renewed Uranium Mining Boom in New Mexico." Author Thomas Power, a specialist in environmental economics, holds that industry claims of potential economic benefit were grossly exaggerated, and "important environmental and social costs" were never considered (2). The report, which might have changed minds had it been considered dispassionately outside of an immediate controversy, looks at past gains and costs from uranium mining in the area and observes that "the role of natural resources in the local economy is *not* diminishing but *changing* from extraction and export to non-consumptive and environmental" (52; emphasis in the original). Power concludes,

> It is vitally important from an economic viewpoint to treat environmental regulation as a part of a region's or state's economic development strategy rather than allowing economic development strategy to constantly undermine efforts to protect social and natural environments. An economic development policy focused on reviving the industries that were important sources of jobs and income in the past by sacrificing additional elements of the natural environment, in the present and future, may actually be *undermining* the economic future of a region. (53; emphasis in the original)

The absence of such discussion has characterized the failed TCP nomination process.

Despite achievements in bonding, bridging, and linkage by Native groups to gain governmental protection for their sacred sites (Forest Service declaration of eligibility for the National Register), the five tribes, their supporters, and the CPRC have yet to find the social resilience at the local level that could allow for the creative change of TCP protection without severe opposition. For that to happen, the Native members of the Mt. Taylor SES would need to meet with the landowner community, perhaps with the assistance of an outside facilitator like the Sonoran Institute, to identify dissensus and to establish common ground, in order to create a mutually beneficial approach to protecting their sacred mountain. If such a coming together does not occur, the crisis surrounding the TCP/mining issue will provide another opportunity for learning from disruption, and may place the Mt. Taylor SES on the verge of another cycle of collapse, reorganization, and renewal.

WORKS CITED

All Indian Pueblo Council. "Companion Resolution for the Protection of Mt. Taylor and All Sacred Sites and Cultural Properties Related to the Pueblos of Acoma and Laguna and the Nineteen Pueblos of New Mexico." All Indian Pueblo Council Resolution 2007–12. 21 Jun. 2007. Web. 5 Jun. 2011.

All Indian Pueblo Council, American Anthropological Association, Association on American Indian Affairs, National Trust for Historic Preservation in the United States, Sierra Club, and Society for American Archeology. Brief of *Amici Curiae* in Support of the Defendants. NM Fifth Judicial District. No. D-506-CV-2009–812. Rayellen Resources, Inc. *et al* v. NM Cultural Property Review Committee. 22 Sept. 2010. Web. 23 Jun. 2011.

Anschuetz, Kurt. "Pueblo of Acoma Significance Statement." Application for Registration: New Mexico State Register of Cultural Properties. Form A, Section 12, 5–36. Revised 18 May 2007. Web. 24 Jun. 2011.

Basso, Keith. *Wisdom Sits in Places: Landscape and Language among the Western Apache.* Albuquerque: U of New Mexico P, 1996. Print.

Benedict, Cynthia Buttery, and Erin Hudson. "Mt. Taylor Cultural Property Determination of Eligibility for the National Register of Historic Places." Mt. Taylor Ranger District, Cibola National Forest, Cibola, McKinley, and Sandoval Counties, New Mexico. Albuquerque, NM: USDA Forest Service, 2008. Web. 20 Jun. 2011.

Benson, Melinda Harm. "Geographies of Mt. Taylor: The Legal, Political and Cultural Implications of Proposed Uranium Mining Development in One of New Mexico's Most Sacred Spaces." Association of American Geographers Annual Meeting. 13 April 2011. Web. 18 June 2011.

Berkes, Fikret, Johan Colding, and Carl Folke, eds. *Navigating Social-Ecological Systems: Building Resilience for Complexity and Change.* Cambridge: Cambridge UP, 2003. Print.

"Challenges to Protecting Mount Taylor." Southwest Research and Information Center: *Voices from the Earth* 11.2 (2010): 4–5. Print.

Ferguson, T. J. "Pueblo of Laguna Significance Statement." Application for Registration: New Mexico State Register of Cultural Properties. Form A, Section 12, 41–76. Revised 18 May 2007. Web. 24 Jun. 2011.

Final Order Approving Nomination for Listing on New Mexico Register of Cultural Properties. New Mexico Cultural Properties Review Committee. 14 Sept. 2009. Web. 10 Jun. 2011.

Goldstein, Bruce E., ed. *Collaborative Resilience: Moving Through Crisis to Opportunity.* Cambridge, MA: MIT P, 2012. Print.

Goldstein, Bruce E., and William H. Butler. "Collaborating for Transformative Resilience: Shared Identity in the U.S. Fire Service Learning Network." *Collaborative Resilience: Moving Through Crisis to Opportunity.* Ed. Bruce E. Goldstein. Cambridge, MA: MIT P, 2012. 339–358. Print.

Hays, Ti. "Court Ruling 'a Setback' for New Mexico's Mt. Taylor." National Trust for Historic Preservation. *Preservation Nation,* 7 Feb. 2011. Web. 20 Jun. 2011.

———. "National Trust for Historic Preservation Names Mount Taylor Near Grants, New Mexico, to Its 2009 List of America's 11 Most Endangered Historic Places." National Trust for Historic Preservation. *Preservation Nation,* 28 Apr. 2009. Web. 21 Jun. 2011.

Hopi Tribe Historic Preservation Office. "Hopi Tribe Significance Statement." Application for Registration: New Mexico State Register of Cultural Properties. Form A, Section 12, 36–41. Revised 18 May 2007. Web. 24 Jun. 2011.

Ingham, Zita. "Landscape, Drama and Dissensus: The Rhetorical Education of Red Lodge, Montana." *Green Culture: Environmental Rhetoric in Contemporary America.* Ed. Carl G. Herndl and Stuart C. Brown. Madison: U of Wisconsin P, 1996. Print.

Jaramillo, Donald. "Judge Reverses Mountain Status." *Cibola Beacon.* 8 Feb. 2011. Web. 13 Aug. 2011.

Kelley, Klara, Harris Francis, and Kurt Anschuetz. "Navajo Nation Significance Statement." Application for Registration: New Mexico State Register of Cultural Properties. Form A, Section 12, 77–93. Revised 18 May 2007. Web. 24 Jun. 2011.

Kirmayer, Laurence J., Megha Sehdev, Rob Whitley, Stéphane F. Dandeneau, and Colette Isaac. "Community Resilience: Models, Metaphors and Measures." *Journal de la Santé Autochtone.* Nov. 2009. Web. 16 Sept. 2011.

Laguna Acoma Coalition for a Safe Environment. "Laguna Acoma Coalition for a Safe Environment (LACSE) Opposes New Uranium Mining on Mt. Taylor Cultural Landscape." Laguna Acoma coalition for a Safe Environment, n.d. Web. 21 June 2011.

Manataka American Indian Council. "Mount Taylor: America's Most Endangered Historic Places." *Sacred Sites.* n.d. Web. 13 Aug. 2011.

MASE Newsletter. Multicultural Alliance for a Safe Environment. Masecoalition. org. 3 Jun. 2011. Web. 15 Jun. 2011.

National Register Bulletin 15. *How to Apply the National Register Criteria for Evaluation.* Washington, DC: U.S. Department of the Interior, National Park Service, 2001. Revised version of 2002. Web. 13 October 2011.

National Register Bulletin 38. *Guidelines for Evaluating and Documenting Traditional Cultural Properties.* Washington, DC: U.S. Department of the Interior, National Park Service, 1990. Web. 13 October 2011.

Navajo Nation Council. "An Act Relating to Resources, and Diné Fundamental Law; Enacting the Diné Natural Resources Protection Act of 2005; Amending Title 18 of the Navajo Nation Code." *Navajocourts.org,* 29 Apr. 2005. Web. 1 Aug. 2011.

New Mexico Cultural Properties Protection Act. NM Statute 18–6A-1 through 6.1978. Web. 15 Dec. 2011.

Noon, Marita. "Public's Best Interest, Disinterested Public." *The Rational Argumentator: A Journal for Western Man* 198 (June 2009). 27 Jun. 2009. Web. 24 Jun. 2011.

Ortiz, Simon J. "Introduction." *Speaking for the Generations: Native Writers on Writing*. Ed. Simon J. Ortiz. Tucson: U Arizona P. 1998, xii. Print.

Paskus, Laura. "Sacred Strife." *High Country News. hcn.org*. 7 Dec. 2009. Web. 18 June 2011.

Pasternak, Judy. *Yellow Dirt: An American Story of a Poisoned Land and a People Betrayed*. New York: Free P, 2010. Print.

Patton, Bruce. "Negotiation." *The Handbook of Dispute Resolution*. Ed. Moffitt, Michael L. and Robert C. Bordone. San Francisco: Jossey-Bass, 2005. 279–303. Print.

Power, Thomas Michael. "An Economic Evaluation of a Renewed Uranium Mining Boom in New Mexico." New Mexico Environmental Law Center: 2008. Web. 14 Aug. 2011.

Sierra Club, Northern New Mexico Group. "Sierra Club, Tribes Team Up to Protect Sacred Lands on Mt. Taylor." *Northern.nmsierraclub.org*, 5 June 2009. Web. 4 June 2011.

State of New Mexico Cultural Properties Review Committee. "Final Order Approving Nomination for Listing on New Mexico Register of Cultural Properties." New Mexico Historic Preservation Division, Department of Cultural Affairs. 14 Sept. 2009. Web. 18 June 2011.

Southwest Research and Information Center. *The Workbook*. 24.3 (1999). Web. 29 Dec. 2011.

Walker, Brian, and David Salt. *Resilience Thinking: Sustaining Ecosystems and People in a Changing World*. Washington, DC: Island P, 2006. Print.

WISE Uranium Project. "New Uranium Mining Projects—New Mexico, USA." *Wise-uranium.org*, 13 May 2011. Web. 4 June 2011.

Zuni Tribal Council. "The Zuni Tribe's Significance Statement." Application for Registration: New Mexico State Register of Cultural Properties. Form A, Section 12, 93–106. Revised 18 May 2007. Web. 24 Jun. 2011.

Part II

Places We Build and Create

5 A Land Ethic for Urban Dwellers

Gesa E. Kirsch

In this chapter, I explore how city dwellers might develop an urban land ethic. Specifically, I argue that we need to develop a more fluid concept, one that includes the different locales in which we might live—past, present, and future—and that pays particular attention to an increasingly mobile society. As Aldo Leopold suggests, "Your true modern [citizen] is separated from the land by many middlemen, and by innumerable physical gadgets. He has no vital relation to it" (223). This claim is particularly true today: we are tethered to technology but not the land. The challenge becomes how to cultivate the "love, respect, and admiration for land" (224) so necessary for a land ethic? As Frank Gaughan notes,

> Despite Leopold's emphasis on local practices and amateur naturalism, the individuals that figure in much of his work seem distant from the life ways of typical city dwellers, whom Leopold laments as a "landless people" ("Land Pathology"). Today, however, we live in an era of landless people, with the majority of the world living in urban areas . . . How, then, shall this growing landless class interpret and apply Leopold's work, so much of which employs examples drawn from agricultural and wilderness settings?

I attempt to answer these questions with reference to an urban area, the city of Boston. Specifically, I analyze the public discourse surrounding the restoration and rehabilitation of the Longfellow Bridge that connects Boston and Cambridge. Plans for the Longfellow Bridge rehabilitation evolved from several community meetings hosted by Massachusetts Department of Transportation (MassDOT) under the Commonwealth's $3 billion Accelerated Bridge Program. According to *Boston Globe* reporter Eric Moskowitz,

> The 21st century reconstruction project proved initially thorny for officials torn between moving as swiftly as possible—and advancing plans to rebuild the bridge as is—and complying more fully with changing state and federal policies aimed at encouraging walking, biking, and transit use for environmental and public health reasons. Those two

goals were at odds because the Longfellow's landmark status meant it could not be widened over the river to provide ample room for everyone.

The state pulled back on its Longfellow plans in 2010 and convened a 36-member task force that included biked, pedestrian, and environmental advocates, neighbors, and civic and business leaders, whose input contributed to the new design. (A8)

Taskforce members were able to articulate a "land, water, air, and transportation ethic" that far exceeds bridge/road design concepts and contributes to an important vision of Boston's future uses of parklands, rivers, forms of transportation, and regional networks of communities. What is interesting about this project is that it is in progress, still being refined through community input, nonprofit organizations, and government agencies, while the early structural work of restoring the Longfellow Bridge has already begun.

The Longfellow Bridge rehabilitation provides a rich example of locales as rhetorical places. As Nedra Reynolds has argued, our perception of places are shaped by language as much as by materiality: rhetoric, metaphor, and imagination all shape how we perceive places, where we visit and linger, where we pass through in a hurry, where we feel safe and invited or uncomfortable and out of place. Discursive constructions of the Longfellow Bridge illustrate the power of language to create places; task force members and residents attending the public meetings described the bridge in many varied terms: as a promenade, a highway, a historical landmark, an access point to the parklands, a commercial artery between two cities, an avenue leading to local hospitals, universities, entertainment venues and businesses, and Boston icon offering beautiful views worthy in their own right. Participants at public information meetings articulated visions that encompass new priorities for the city, such as reclaiming parkland, restoring the Charles River shoreline, and promoting alternative modes of transportation.

Those visions resonate with Aldo Leopold's call for a land ethic, albeit a land ethic in an urban context. Leopold suggests that we "quit thinking about decent land use solely as an economic problem" (224). Instead, he proposes that we "examine each question in terms of what is ethically and esthetically right, as well as economically expedient" (224). Leopold, of course, was concerned about the integrity of ecological systems or what he calls "biotic communities," where each member—whether plant, animal, or microbe—has a right to existence, regardless of economic value. He did not concern himself with urban parklands or bridge design. Yet his call for aesthetic appreciation and ethical thinking helps to explain the passion, concern, and arguments articulated by city residents and task force members.

Whereas Leopold suggests that "your true modern [citizen] is separated from the land by many middlemen, and by innumerable physical gadgets" (223), I argue that city dwellers can often be fiercely proud and protective of their parks, water fronts, lakes, and rivers, even if the land was created artificially, as is the case with the Esplanade parkland. In other words, many of the city residents who participated in the public meetings of the Longfellow Bridge rehabilitation understood deeply, as did Leopold, that "we can be ethical only in relation to something we can see, feel, understand, love, and otherwise have faith in" (214). They also understood that we need to treasure the little bits of nature around us and fall in love with the world through observation, spending time outdoors, and having fun, a point raised by David Gessner who chronicles the recent extension of pedestrian paths along the Charles River. This development, Gessner notes, has increased access to the river shore line for many residents who now claim a strong stake in the Charles River's health, beauty, and protection (169).[1]

This case study also addresses, in part, a question raised by Elenore Long about the role of local knowledge in environmental rhetoric and civic action:

> As different as the rhetoric of environmentalism and urban community action seem to be, they speak of the field's growing interest in activism and public engagement. They are also struggling with a shared couple of questions: how to elicit and validate local knowledge along discourses—whether discourses of policy, science or bureaucracy—that tend to dismiss it, and how to combine local knowledge with impact that is at once transformative and sustainable (13).

In my analysis, I set out to contribute to the understanding of environmental rhetoric through a specific case study. I focus on two public meetings hosted by the MassDOT: the October 6, 2010 meeting that presented the recommendations of the Longfellow Bridge Rehabilitation Task Force and invited public comments, and the April 11, 2011 meeting that provided a progress report on the early rehabilitation work and invited further public comment.[2] In this way, I anchor my argument in a specific time and place, and "contribute to the telling of stories of actual events and the people who participate in those events" (Goggin 5) as they shape public discourse and life of a city. In so doing, I am fully aware that I tell my own story, as a resident of Boston and as someone deeply interested in environmental discourse and sustainability.

The reason I became interested in the Longfellow Bridge restoration is that I cross it almost daily, whether on the redline subway, on bicycle or on foot and occasionally by car. As a runner, I am often forced to negotiate a very narrow Boston-bound sidewalk (four feet wide) with pedestrians, strollers, and other runners, whereas cars enjoy fifteen-feet wide traffic lanes (the standard width more common for federal highways, not city streets). When a meeting was scheduled for the presentation of Taskforce recommendations, I went with a friend, a cyclist, so that the voices of at least one runner and one cyclist would be heard. In the span of the two-hour meeting, I was surprised to hear so many different stakeholders express their concern about the Longfellow Bridge and the surrounding environment:

there were many other runners and cyclists, but also citizens from many other walks of life who claimed a stake not only in the bridge—the sidewalks, bike lanes, traffic lanes and train tracks—but also in the parkland and river beneath the bridge, in the access points on and off the bridge, and they argued for promoting healthier life styles and imagining a vibrant future that makes sustainability and livable communities a priority.

The case of the Longfellow Bridge rehabilitation shows that local knowledge can influence environmental rhetoric profoundly. In fact, it was the voices of members of the taskforce—representing many diverse stakeholders—and of those attending the public meetings that changed the agenda from an engineering and traffic management project to one of urban planning and environmental activism.

> The Longfellow Bridge (originally, the Cambridge Bridge) is one of the most architecturally distinguished bridges in Massachusetts. Located on the site of the 1793 West Boston Bridge, this graceful steel and granite structure was completed in 1908, and renamed in honor of Henry Wadsworth Longfellow in 1927. The bridge joins . . . Boston with . . . Cambridge and carries approximately 90,000 transit users [on the

Figure 5.1   View of the Longfellow Bridge from the Prudential Tower observatory (Strong).

redline subway], 25,000 motor vehicles, and a significant number of pedestrians and bicyclists each day. The two central piers carry the signature pairs of neoclassically inspired . . . granite towers that have given the bridge its popular nickname—the Salt and Pepper Bridge. (MassDOT "FAQ Handout")

Figure 5.2 The Piers (Salt and Pepper Shakers). Longfellow Bridge between Cambridge and Boston, Massachusetts, USA (Daderot).

Initially, the project started with the impetus to fix the bridge before it would fall down, or, to quote MassDOT, "a primary objective of the proposed rehabilitation and restoration of the Longfellow Bridge is to address the bridge's current structural deficiencies, upgrade its structural capacity, and bring the bridge up to modern codes" ("FAQ Handout"). Yet the project became much more complex: after studying the bridge and its role in the surrounding communities for more than six months, the taskforce articulated a vision not only for the rehabilitation of the bridge, but for the uses of the parkland, the river, and the future of urban living in Boston and Cambridge.

Task Force members represented an astonishing thirty-six stakeholders who participated in the process of drafting recommendations; they include neighborhood organizations (Downtown North, Kendall Square, West End, and Beacon Hill Civic Associations); business associations; environmental and parkland associations (Esplanade Association, Charles River Conservancy, Charles River Watershed Association), river users (Community Boating, Riverside Boat Club), pedestrian and bicycle groups (Walk Boston, Mass Bike, Boston Cyclist Union), history preservation groups (Boston Preservation Alliance, Mass Historic Commission), universities and hospitals (MIT, Mass Eye and Ear, MGH), representatives of the city governments of Boston and Cambridge, representatives the state government, the Federal Highway Administrations, and groups invested in reimaging urban living (Livable Streets, A Better City).

At the public meeting on October 6, 2010 during which the task force presented its recommendations (three distinct alternatives) and sought public feedback, State Representative Marta M. Walz, who represents constituents in both Boston and Cambridge, encouraged bold thinking, courage, and imagination in planning the future:

> Our challenge is to take this 19th century bridge—it was modified in the 20th century in a car-centric culture—and transform it into how we travel today and how we will travel across this bridge for the balance of the 21st century. And that is an enormously difficult challenge . . . I urge all of us to be bold, to be visionary, and design a bridge that shows that we will not be bound by old ideas and old ways of thinking. (18–19)

Adding a touch of humor, Walz noted, "The Longfellow Bridge is one of my oldest and neediest constituents, and one that doesn't complain very much. And so since it can't speak for itself, I'm glad that there are so many of you here tonight to help speak on behalf of the bridge and what it can and should be for all of us" (17).

The constituents rose to Walz's challenge: they expressed their appreciation, concern, even love for the local, urban parkland when they spoke at public meetings to discuss the restoration of the Longfellow Bridge. A close analysis of the public discourse surrounding the bridge rehabilitation shows how specific locales can become "rhetorical places for examining complexities and discursive constructions of change, resilience, and sustainability" (Goggin "Call for Proposals"). I use the broad framework of "change, resilience, and sustainability" to group the concerns voiced by different stakeholders, noting that many concerns overlap and shift from one category to another. Nevertheless, these categories are useful in grouping the interests of different stakeholders.

ARTICULATING RESILIENCE AND PRESERVING A LANDMARK

Some task force members argued for restoring the original bridge design that provided easy connections to the parklands and river, and for keeping key components of the bridge within an historical context and historical intent. They noted that the bridge has become an icon of the city, as its nickname, the Salt and Pepper Bridge, suggests. They spoke passionately of the bridge's beautiful design and the scenic views from the bridge deck. In other words, they represented the bridge as an important Boston landmark, as a historical structure that needs to be restored to its early glory, both in function—easy access to parkland and the Charles River—and in form—as an aesthetic object, a place that invokes beauty, history, and scenic views.

Herb Nolan of the Esplanade Association reminded the community of the historic nature of the bridge and the connection it used to provide to the parklands:

> It's remarkable how far we've come in terms of erasing the historic function of this bridge, 100 years ago when it was built. It was intricately connected to the parkland on both sides. It was designed to do that. And until six months ago, there wasn't much consideration being given to those vital connections . . . Today we all acknowledge that that's a core part of the purpose and need of this bridge. It was there historically. So, when we restore the Longfellow Bridge, we are restoring not only its structure and its materials, but its core function, which is connectivity. (MassDOT, "Task Force" 34–35)

John Shields, also of the Esplanade Association, argued for the historical significance of the parkland, of reclaiming parkland whenever possible, of imagining what the parkland could be:

> Things go under the bridge. And, boats go under the bridge. People go under the bridge also. There are two arches on the landside, on the Boston side. And we want to go on record now as part of the vision that we want one of those arches back as parkland . . . We feel like all that traffic that is on Storrow now could go under the most inbound part [arch]. (MassDOT, "Task Force" 85)

Kevin Wolfson, of Livable Streets and a cyclist, spoke to the beautiful views from the bridge:

> I think one of the features of the Longfellow Bridge that no one tonight has mentioned . . . is the incredible view . . . It's something that I didn't really recognize by living in the area and riding my bike into and out of the city when I was a kid, until I moved back here after college. And I didn't recognize it because the Longfellow Bridge is so uninviting to this point.
>
> I think [we should] invite people onto the bridge . . . to enjoy that view, really recognize it for what it is and to see it as a feature of Boston . . . It would have been nice as a kid in Boston or in the area to have seen that bridge instead of having to wait till I was an adult. (Mass-DOT, "Task Force" 87–88)

Referring to the beautiful views from the bridge, Tom Nally, from a Better City, proposed that "the sidewalk should be a wide promenade of approximately 15-foot six-inches wide" (MassDOT, "Task Force" 69).

These speakers valued the historic and aesthetic nature of the bridge, its role in allowing easy access to the parkland and river, and the beautiful views from the bridge. They also mentioned numerous times the role of the bridge as an icon of the city and as a gateway to Boston and Cambridge. We can hear Leopold's words reflected in these residents' remarks: "It is inconceivable to me that an ethical relation to land can exist without love, respect, and admiration for land, and a high regard for its value" (223). Residents' comments illustrate a love, indeed a passion, for the Longfellow Bridge and its surrounding environment, and in fact, it is this vision of the bridge as an integral part of Boston's identity and environment that gave rise to the broad range of task force recommendations.

Task force members viewed the restoration project as a chance to connect the past with the future and emphasize historic intent and aesthetic qualities in addition to the more mundane tasks of reinforcing the bridge's steel structure and solving traffic problems. They could also be said to "reveal" a sense of place, as Dan Shilling explains, because "in attempting to uncover a sense of place, *distinction* is key" (8). Task force members and residents clearly recognize what is unique and distinct about Boston as a locale, and, working "in concert with a community's natural and cultural heritage" (6), they aim to restore and preserve that unique place.

ADVOCATING SUSTAINABILITY AND HEALTHY COMMUNITIES

Arguments also turned to building sustainable communities and developing healthier lifestyles. Residents asked questions including the following: Does the Longfellow Bridge encourage people's connection and access to waterfront and parkland? Does it encourage different modes of transportation? Do different users of the bridge feel safe bicycling, walking, running, strolling across the bridge, perhaps even with family in tow? Or, as Jeremy Mendelsohn asked, does it feel like a highway, where cars are the default mode of transportation, and other uses are left to "fight over the scraps" (105)? Peter Stidman of Boston Cyclists Union argued for

> encourag[ing] families to use this bridge, riding with their kids. It connects the Esplanade to the other side of Memorial Drive. It's an historic bridge. It's a beautiful bridge. I think there's a lot of potential there that would be missed if we went with a five- or six-foot wide bike lane that really is only comfortable for commuters and people that are experienced with riding in traffic. (MassDOT, "Task Force" 102)

Tara Kruger, a resident who uses all modes of transportation added,

> This is really an ideological problem . . . for 70 years, for three-quarters of a century, we've been subsidizing driving. We've given an enormous subsidy to driving for an incredibly long time, and now is the time to subsidize alternative modes of transportation that will help us get to where we want to be as a society, that will help combat health problems, and that will support livable communities all over the place. (MassDOT, "Task Force" 111)

Kruger's statement was followed by applause from the group.

Discussion also turned to environmental integrity of parkland, shoreline, and water and air quality. The Charles River Watershed Association (CRWA) articulated concerns about storm water runoff and the health of the river. The Esplanade Association (TEA) picked up the sustainability argument with concerns about restoring more of the original parkland, now used as parking lots, in order to help meet two goals: one, provide better storm water runoff, and two, provide better access to parkland from Massachusetts General Hospital (MGH), an argument based on both the health of the land and river, as well as the health of park users.[3] Margaret Van Duesen of the Charles River Watershed Association explained,

> I want to bring up the storm water management problems off the bridge, which are quite challenging into the parkland and to meet regulatory requirements. And to echo Herb Nolan's comment about the potential for using what is DCR owned land, [now used] for some of the Mass Eye and Ear parking lot, for managing that storm water. Because it really could be a win/win in terms of restoring some parkland and access, and also [in terms] of constructing a storm water park that can handle the run off on at least the Boston side. (MassDOT, "Public Information" 70)

Paul Greenfield, a driver, expressed concern about air quality:

> I am speaking as one of the 28,000 who drives. I live in town, but I do drive over the bridge . . . Any change would be admirable, because we do need to promote more walking and bicycles, but the fact is there will be a queue, which means there will be more emissions and more idling. (MassDOT, "Task Force" 86)

An astute observer noticed the importance of access to the Esplanade for MGH staff, patients, and visitors made another interesting comment about parkland access and healing. Herb Nolan stated,

> I was just talking to some staff members at MGH the other day about access to parks, how important it is to the patients, as well as the visitors. It's incredibly dense world over there, people with very little room and very little access to open space. (MassDOT, "Public Information" 62)

Here speakers and taskforce members expressed concern about what Leopold calls the "biotic community" in which each member has a right to existence, regardless of economic value: "a land ethic changes the role of *Homo sapiens* from conqueror of the land community to plain member and citizen of it. It implies respect for his fellow-members, and also respect for the community as such" (204). During the community meetings, residents spoke passionately about the importance of the health of the river, the health of the land and the shoreline, the health of the human community, and the healing properties associated with having access to beautiful, natural places.

The case of the Longfellow Bridge restoration, I want to suggest, represents a striking example of how a project can evolve from what was once considered an engineering project to a community visioning project in which citizens reclaim their right to speak and assert that local knowledge is relevant to the design and rehabilitation of the historic bridge. Residents and task force members became rhetorical agents actively engaged in envisioning a common future, in reclaiming ownership in the public process, and in constructing locales as rhetorical places. They viewed the bridge, as Seth Zeren proposes in a personal interview, as a "living entity" leaving room for growth and accounting for changes in use

In other words, rhetorical constructions of place are powerful agents that inform contexts for sustainable design, not just bridge design, and perhaps more importantly, construct places that people identify with, claim an interest in, appreciate aesthetically and perhaps come to know and love. In short, rhetorical constructions of place might become a powerful factor in developing a land ethic for urban dwellers.

ARGUING FOR CHANGE AND IMAGINING THE FUTURE

A number of task force members and residents participating in the April 11, 2011 MassDOT public information meeting expressed their concern for the long-term decisions about to be made: how bridge design would need to accommodate shifts in mode of transportation (e.g., more public transit users, more cyclists), but even more importantly, how bridge design would enable—or limit—different forms of transportation. Ken Kreckameyer of Livable Streets noted,

> The world is changing so significantly that we ought to be doing designs that are very bicycle oriented and very pedestrian oriented and does, in fact, control the cars at a slow speed so that in, in fact, the people are using it [the bridge] are safe. (53)

Charlie Denison spoke to the importance of installing the crash barrier that separates the sidewalk from the traffic lanes at just the right spot:

> Fifty to seventy years out, trends of bikes and pedestrians numbers are increasing; motor vehicle volumes are decreasing; getting that barrier right today is absolutely key. (MassDOT "Task Force" 76)

John Collins added,

> I would point out that we're not only talking bicycles, pedestrians and conventional motor vehicles, we can expect an increase . . . in unconventional human power vehicles, and also small motorized vehicles; mopeds, motorized bicycles, electrically powered bicycles. (MassDOT, "Public Information" 67–68)

What is striking about these arguments is that they concern the future: they provide a vision that encourages, even demands change (e.g., new modes of transportation; attention to increased population, more concern for healthy and sustainable communities). Several speakers focused on anticipating future generations of bridge users; they were aware of the challenges of imagining the future of the city and the future of different forms of transportation. They were also humbled by the fact that the group of stakeholders and concerned citizens assembled at this meeting would influence decisions that may have long-term, perhaps unintended consequences, and that they, as city residents, might be judged by future generations according to those decisions. Steve Miller noted,

> We're building this not even for ourselves. We're building this for our grandchildren. It's amazing. For some of us, our kids are grown, our

kids will be dead while this bridge is still here. So this is a critical link to the future we're playing with. (MassDOT, "Task Force" 45)

Tom Nally further explained,

It's important to address the needs of today while building for the future conditions and opportunities. The cross-section of the Longfellow Bridge that we start to build next year must accommodate both today's needs and those that will evolve over the next 50 years or more. (MassDOT, "Task Force" 69)

Fred Salvucci concluded,

The use of this bridge is going to change a lot over the estimated 75-year life that you're designing for this round. So it's important that there'd be flexibility because we all make mistakes in guessing what our future is going to be like. The key decision that has to be made now, if you're going to get to fix this bridge before it ends up in the river, is to decide where the [crash] barrier . . . goes. (MassDOT. "Public Information" 76–77)

Clearly, community members invoked a sense of ethics, although that is not the language they used to articulate their hopes and visions for the future. By accepting responsibility to future generations, these speakers imply—as does Leopold—that the long-term vision matters (not only the short term economic gain); that we owe it to those not yet born to be thoughtful, caring, and flexible as we imagine the future modes of transportation, leaving room for the yet-to-be-invented ways of traveling and living.

READING AGAINST THE GRAIN OF HISTORY

Lest this account of public discourse and environmental debate reads too much like a progressive dream, it is important to note that Boston is not an "ideal" place where citizens engage in environmental rhetoric and dream big—although, as the sample quotes above illustrate, Bostonians do love to dream big. Rather, it is critical to read this case against the grain of history. Boston has a long history of ignoring citizens' voices and their visions for urban spaces. Perhaps the most notorious case of a clear violation of citizens' trust in the public process was the complete demolition of an ethnically diverse, working-class, neighborhood called the West End. In the late 1950s, during the height of the "urban renewal era," the city seized an entire neighborhood through eminent domain law, demolished the homes of thousands of residents, and built "superblocks," now filled with high-rise residential towers, government center plaza, and city hall.[4] Equally notorious was the building of Storrow Drive, a highway that runs along the Esplanade. In 1949, the state legislation voted to approve a highway to be built on Esplanade parkland, thereby ignoring citizens' voices and the estate of the Storrow family, who had given money and land intended to enhance the Esplanade embankment.[5]

Distrust in Boston authorities and government processes continue to this day, as evidenced in the transcripts from the April 11, 2011 public information meeting. Several attendees expressed dismay that six months after the task force had presented its recommendations and invited public comment, MassDOT was not ready to present its preferred alternatives, and more critically, announced that it would submit its preferred alternatives to the Federal Highway Administration for an environmental assessment review without further vetting by Boston and Cambridge residents. As Ken Kreckameyer put it,

> It seems to me that it would be appropriate for the state to be sharing with citizens of this state, the information about what the preferred alternative is from the part of the highway department and the government, long before we hear from Federal Highway [Administration]" . . . It seems as though you're asking us to be complicit in a process that I suspect many of us have no faith in at the moment. (51–52)

Another issue that is critical to note here is wealth stratification in the city. The neighborhoods most directly affected by the Longfellow Bridge Rehabilitation are some of the wealthiest in the city—the Back Bay, Beacon Hill, West End, and, on the Cambridge side, Kendall Square. The Esplanade abuts the Back Bay, an area of concentrated wealth and power. Those who participate in public meetings are those who typically live in wealthier neighborhoods and are more likely to have the money and influence to make their voices heard and have government officials respond to their concerns.

Yet, the passionate engagement that emerged from the public meetings reveals that city residents do value the parkland, river, the air and water quality, the bridge as an icon, a landmark, and a beacon pointing toward a brighter future. In many ways, these citizens were able to articulate a water, air, and land ethic that is forcing policy makers, state and city government, and MassDOT to rethink the project in much broader terms than they had originally anticipated doing. Of course, it remains to be seen how exactly the Longfellow Bridge restoration and the future of the cities of Boston and Cambridge will unfold. The federal Environmental Assessment was released for public comment in February 2012; Phase 2 of the Rehabilitation is scheduled to begin in 2013; and final completion of the project is anticipated in late 2016. The bold visions articulated by taskforce members and citizens remain visions at this point; they could change, realign, derail, or move forward. Only time will tell.

ACKNOWLEDGMENT

I would like to thank Ryan Miamis for his excellent research assistance with many of the details pertaining to the Longfellow Bridge Restoration project. He is an outstanding collaborator, researcher, and interviewer who provided me with valuable insights and feedback many times.[6]

NOTES

1. Gessner explains, "It is not the nature of the national park, the nature you need to drive or fly to see. Instead, it's the nature of the creek that runs through your neighborhood, the nature of the abandoned lot, the nature of the small secret patch of beach protected by rocks. I understand that there are those who would scoff at my trying to make claims of wilderness in Needham [a suburb southwest of Boston]. But I think we are making a deadly mistake if we ignore the smaller, more compromised patches, since that is what so many of us are left with" (101).
2. Transcripts for all public meetings, handouts, and slide presentations are available at the Massachusetts Department of Transportation Website.
3. Mass General Hospital is one of the major hospitals in Boston, world-renowned in many medical specialties, and occupies a staff of several thousands. Currently access from MGH to the park and river is very challenging. One needs to cross many lanes of traffic before being able to reach a pedestrian bridge that leads across Storrow Drive to the Esplanade. This route is very uninviting; hence very few of the thousands of MGH staff, patients and visitors ever visit to the Esplanade.
4. Bostonians did not forget the brute force of power that razed their neighborhood. Today, there is a West End museum with rotating exhibits showing the early neighborhood; a West End neighborhood association that is reclaiming its past identity; and an annual reunion mass, called the "West End mass," held for displaced residents in the West End church, one of the few remaining buildings from the neighborhood.
5. A lot of parkland was lost on both sides of the bridge and the staircases that allowed easy access to the parkland were cut off when Memorial Drive and Storrow Drive were built as parkways during the early 1950s. According to Wikipedia, "the parkway is named for James J. Storrow, an investment banker who led a campaign to create the Charles River Basin and preserve and improve the riverbanks as a public park. He had never advocated a parkway beside the river, and his widow publicly opposed it."
6. Thanks also go to members of my writing group—Tina Bacci, Rosaleen Greene-Smith, and Jim Webber—for feedback on an early draft of this chapter.

WORKS CITED

Daderot. *Longfellow Bridge Between Cambridge and Boston, Massachusetts, USA. Wikimedia Commons*, 26 March 2011. JPEG file.

Gaughan, Frank. "Call for Proposals: ASEH Panel /Paper Proposal Aldo Leopold in Urban Settings." List-Serv posting. 28 June 2011. Web. 16 Dec. 2011.

Gessner, David. *My Green Manifesto: Down the Charles River in Pursuit of a New Environmentalism*. Minneapolis, MN: Milkweed, 2010. Print.

Goggin, Peter. "Introduction." *Rhetorics, Literacies, and Narratives of Sustainability*. Ed. Peter Goggin. New York, Routledge, 2009. 1–12. Print.

———. "Call for Proposals for a collection, *Environmental Rhetoric: Ecologies of Place*." Flyer distributed at the Conference on College Composition and Communication, April 2011.

Leopold, Aldo. *A Sand County Almanac and Sketches Here and There*. 1949. Intro. Robert Finch. Special Commemorative Edition. New York: Oxford UP, 1989. Print.

Long, Elenore. "Rhetorical Techne, Local Knowledge, and Challenges in Contemporary Activism." *Rhetorics, Literacies, and Narratives of Sustainability*. Ed. Peter Goggin. New York, Routledge, 2009. 13–38. Print.

MassDOT (Massachusetts Department of Transportation). Longfellow Bridge Rehabilitation and Restoration Project. FAQ Handout. 28 Sept. 2010.

———. Task Force Public Information Meeting Transcript, 6 October 2010. Web. 20 Jan. 2012. http://web.massdot.net/CharlesRiverBridges/LongfellowBridge-Documents.html

———. Longfellow Bridge Rehabilitation Task Force. Draft Recommendations. 6 Oct. 2010. Web. 20 Jan. 2012.

———. Public Information Meeting Transcript, April 11, 2011. Web. 20 Jan. 2012.

———. Longfellow Bridge Rehabilitation Project: Bridging the Past and the Future. Project Update Handout. 11 Apr. 2011.

Moskowitz, Eric. "Longfellow Loses Outbound Car Lane in New Design." *Boston Globe*. 8 Feb. 2012. A1+.

Mott, Meg. "Thinking Like a Person of Color: What the African-American Experience Offers the Land Ethic." Paper prepared for presentation at the American Society for Environmental History, Panel on "Land Ethics for the Landless: Refiguring Aldo Leopold for the Urban Age," Madison, WI. March 2012.

Reynolds, Nedra. *Geographies of Writing: Inhabiting Places and Encountering Difference*. Carbondale: Southern Illinois UP, 2004. Print.

Shilling, Dan. "Sense of Place and the Triple Bottom Line." Manuscript, September 20, 2011, for *Rethinking Sustainability: Multicultural Land Ethics and the Global Search for Quality of Life*. Ed. Bill Forbes and Teressa Trusty. Forthcoming.

"Storrow Drive." *Wikipedia: The Free Encyclopedia*. 20 Jan. 2012.

Strong, Larry. *View of the Longfellow Bridge from the Prudential Tower Observatory*. *Wikimedia Commons*, February 2005. JPEG file.

Zeren, Seth. "Towards an Urban Land Ethic." Unpublished manuscript. May 2010.

——— "Personal Interview." Conducted by Ryan Miamis. Newton, MA. 8 Dec. 2010.

6 "We Face East"

The Narragansett Dawn and Ecocentric Discourses of Identity and Justice

Matthew Ortoleva

Native American environmental advocacy before the 1960s and the emergence of the direct-action oriented American Indian Movement is often overlooked. However, prior to the 1960s and the explosion of the contemporary environmental movement, Native American environmentalism simmered below the currents of the dominant discourses of American society, often in forums that are today difficult but necessary to locate, as these little known artifacts demonstrate a historical commitment to place. Native American environmentalism is marked by beliefs, behaviors, and an ecological ethic rooted in a deep spiritual connection to the natural world, often with particular places or natural phenomena taking on sacred meaning. Moreover, Native American environmentalism is grounded in relationships to the North American hemisphere that can be traced back generations, creating a living land trust that is thousands of years old (Grinde and Johansen 1). One important historical and easily overlooked forum for Native American environmentalism is *The Narragansett Dawn*, a magazine published during the mid-1930s by the Narragansett people of southern New England for the purpose of providing a space for the Native American voice without first being filtered by Anglo-European gatekeepers.

For the Narragansett, *The Narragansett Dawn* served as the forum for the creation of what Nancy Fraser calls a "subaltern counterpublic," which is a discursive space where "members of subordinated social groups invent and circulate counterdiscourses to formulate oppositional interpretations of their identities, interests, and needs" (67). As Frazer so aptly points out, the public sphere is a space of conflict, imbued with power structures and saturated with domination and subordination. The discursive and expressive norms of dominant groups are privileged, and without discursive assimilation, subordinate groups are often locked out. As such, counterpublics emerge "in response to exclusions within the dominant publics" (67) and out of demand to speak in one's own voice by "expressing one's cultural identity through idiom and style" (69). During the 1930s, the Narragansett needed to create a discursive space for the enactment of an *eco-centric* voice that had been shaped over generations of close living to the Narragansett Bay Watershed.[1] The first editor of *The Narragansett Dawn*, Princess Red

Wing, articulates the strong sentiment behind the need for this counterpublic space in the first issue: "No white person can read the heart of the Indian as can a son or daughter of the Red Man's own race. Judge these pages from the Red Man's views. These columns come not from the experienced pens of journalists, but from the hearts and firesides of Narragansett Indians"[2] (2). The identity and *ethos* that Princess Red Wing begins to construct for the Narragansett is one of an embodied relationship and closeness to Narragansett Bay invoked by the appeal to hearth, heart, and home.

Prior to Princess Red Wing's editorial there was very little space for the voice of the Narragansett, and according to anthropologist, Paul Robinson, by the 1930s a history of cultural and physical genocide, severe economic depression, and a systematic campaign of detribalization by the State of Rhode Island fueled by the Dawes Act of 1887 had taken its toll on this proud people. Under the Dawes Act and the foundational assumption that private property ownership would lead Native Americans to cultural assimilation, all but two acres of the ancestral reservation lands of the Narragansett were sold to the State of Rhode Island for a total of $5,000, with each individual share totaling $15.43 and private property allotments made available for individuals, essentially ending tribal ownership of land and recognition of the Narragansett as a distinct tribe[3] (85). Robinson suggests that historical records of the time indicate a fair and open process for the sale of the land and that most Narragansett saw the sale as equitable (85). However, Robinson also points out that Narragansett cultural historians dispute these historical records and counter these claims by suggesting that tribal council members at the time were subject to threats from both outside and inside the tribal community, and that after the vote to sell the land several Narragansett refused to collect their shares (85). Raw feelings of this time are evident in the first issue of *The Narragansett Dawn*, where Princess Red Wing charges that despite the fact that the State of Rhode Island "recorded the tribe extinguished" the Narragansett persevere (2). For Princess Red Wing, this act of recording the tribe extinct—a rhetorical act of domination—is not enough to destroy the embodied connection that the Narragansett have with this place; cultural existence, for her and other Narragansett, was an important and immutable truth.

During the 1930s, the Narragansett Tribe was being motivated by a change of mood at both the local level and the national level. On the local level, as indicated in the June 1935 issue of *The Narragansett Dawn*, the Narragansett held its first tribal meeting in fifty-three years (24), demonstrating that a new sense of solidarity was taking hold among the Narragansett people. At the same time, the Narragansett joined in a national sense of hope that was beginning to take shape around Roosevelt's New Deal policies. In the first issue of *The Narragansett Dawn*, Princess Red Wing references the New Deal in her explanation of the name of the publication: "We have called this monthly booklet, 'The Narragansett Dawn' because we are watching for better times in the '*New Deal*' with our fellow countrymen" (2; emphasis in original). The Narragansett's collective identity, which the editors and contributors of the monthly magazine envisioned, would be rooted in traditional cultural values and beliefs; a strong sense of self-reliance and pragmatism; and a deep-seeded spirituality, which had been reshaped over two centuries by the weight and influence of Christianity. The fusion of Christianity and traditional Narragansett spirituality, rooted in a profound connection to place, would serve as a powerful trope in the formation of the Narragansett's counterpublic. This fusion, articulated in the pages of *The Narragansett Dawn*, offered a foundation for the Narragansett's efforts to reestablish a collective identity and once again assert an eco-centric worldview that would serve as the cornerstone for the tribe's commitment to its ancestral home.

ECO-CENTRISM AND NARRAGANSETT IDENTITY

Robert Kern explains that eco-centrism is a heightened understanding of the world beyond ourselves and an ethico-philosophical perspective with attunement to place (426). In the same vein, political theorist Robyn Eckersley argues that "[e]cocentrism is based on an ecologically informed philosophy of *internal relatedness*, according to which all organisms are not simply interrelated with their environment but also constituted by those very environmental interrelationships" (49). Rhetorical inquiry into eco-centrism means inquiry into the discourses that both shape and manifest from this ethico-philosophical perspective. As Christian Weisser suggests, "Through discourse, we negotiate our relationship with other people, things and places, and through both internal and external dialogue (writing, speaking, and thinking) we establish who we are and how we fit into the world" (90). The core of an eco-centric discourse, however nuanced and shaped by specific context, rests in "a sense that we share our place with all that is other-than-human within it" (Kern 426). An eco-centric discourse, like all discourses, creates and authorizes worldviews, attitudes, and behaviors; serves as support for claims and arguments; and serves as the basis for identity formation. Accordingly, if we accept John Paul Gee's suggestion that discourses are identity kits (7), then we accept that eco-centric discourse will help build, in both individuals and collectives, what Mitchell Thomashow calls an *ecological identity* (3). For Thomashow, "ecological identity refers to all the different ways people construe themselves in relationship to the earth as manifested in personality, values, actions, and sense of self" (3). As such, eco-centric discourses and acts of rhetoric can construct an ecological identity or affirm an already well-rooted identity.

The ecological identity of the Narragansett, whether individually or collectively, is rooted in a deep connection to Narragansett Bay and its surrounding area, and this ecological identity is often expressed with specific references to Narragansett Bay, or by demonstrating a well-developed local knowledge of what it means to live in and, indeed, be born of a Northeast woodland and coastal ecological community. Consider, for example, the writings of Lone Wolf, a Narragansett and frequent contributor to *The Narragansett Dawn*. Lone Wolf's writing exemplifies the eco-centric discourse that is interwoven into every issue of *The Narragansett Dawn*. In one particular article, from July 1935, Lone Wolf writes about an encounter with a partridge that he had while fishing for trout one spring. Lone Wolf's story of his encounter offers a clear sense of attunement to place:

> I saw a partridge fly up near a pile of dry brushes, beside my path, near the stream. Noticing this was a good place for a nest, because of the brush, and old down timbers and hollow stumps. I quietly went on fishing. Soon I heard a noise that started with a loud thump, was repeated six to ten times, stopping at intervals for about a minute. This was the cock-partridge drumming. He does this the same as a rooster crows and you can hear the reverberation about five hundred yards, without detecting the position of the partridge. I left off fishing, crept along the path, very carefully and softly. If I had made a sound, the show would have been over. There in a little opening, on a hollow log, was a cock-partridge, beautiful to behold . . . watching, strangely fascinated, I moved for a better view, but unfortunately stepped on a dry stick which cracked and there was a whirl of wings. The show of the wild was over. At the full of the moon the partridge will drum, at intervals all night, except in winter. You know, in winter he roosts up in the pines next to the trunk out of the wind and storms. ("Encounter" 18)

Lone Wolf ends his narrative by suggesting the traditional way to catch a partridge is to go out at sunset with a stick with a loop on it and snare the partridge off its roost ("Encounter" 18). This personal story shows more than just a passing encounter. Lone Wolf calls on his direct experience of this particular place to be as unobtrusive as possible while still observing the natural processes that are unfolding. Note also the references to the seasons. In another issue of the publication, Lone Wolf writes about the importance of fishing in season to protect the health of the various species that are found in New England waters ("Trout" 27). Historically, the Narragansett relied on local knowledge of the way their ecological community changed from season to season for reasons of harvesting, preserving food, and moving between a home in the shelter of the forests and a home by the waters of the bay. Lone Wolf demonstrates how the Narragansett, like the partridge, shape behavior around the seasonal rhythms of the ecological community. Lone Wolf feels a connection to his ecological community and demonstrates an understanding of the behavior and life cycle of one of its other inhabitants, the partridge. His story is one of appreciation bordering on awe—evidenced when he calls the partridge "beautiful to behold," as well as one of practical interrelationship—as the beautiful partridge becomes a potential food source.

In another article titled "Little Things in Nature," in the April 1936 issue of *The Narragansett Dawn*, Lone Wolf once again conveys a story that demonstrates a sense of place attunement and understanding of ecological relationships. After mentioning a season filled with "a great deal of rain" and "much melting snow," Lone Wolf reflects on a walk in the swamp where he usually sees small game (285). However, on this particular trip Lone Wolf "saw none" (285). The absence of small game, the reader learns, is because "Nature has a way of telling her little ones to look out for floods" (285). According to Lone Wolf, with deep snow and extended cold through February, "when a warm spell came in March, the thaw came quick. The swamp, a foot above the water level, was over night about five feet under water" (286). Lone Wolf goes on to suggest that the small game was attuned to the changes in season, and he was attuned to the habits of the small game. He writes of the coming thaw and the ensuing flood:

> The game knew this. They took the hint early and left the swamp. I was cutting wood there and took my lesson from the little ones. I stopped cutting and drew the wood out and the next day the place was a lake. If I had not noticed the small game's departure I would have lost my wood which meant a great deal to me. (286)

Lone Wolf shows such a connection to his ecological community that he draws warning from nonhuman inhabitants that live there. He inhabits the same place, draws wood for warmth from this place, and relies on animals for lessons of caution, all while staying attentive to the rhythms of the natural world.

Evident in Lone Wolf's writing is a strong sense of internal relatedness, which, as Eckersley suggests, is the foundation of an eco-centric perspective. Lone Wolf writes himself as part of the biotic community, resisting the view that the non-human world is simply a backdrop for the activities of human agents. Lone Wolf demonstrates that an eco-centric perspective recognizes that the different members of the nonhuman community are also appreciated as important in their own terms and having their own modes of being (Eckersley 55). Lone Wolf's day-to-day interactions with these other creatures who share his ecological community constitute his sense of belonging, a belonging that is intertwined with self-reliance and pragmatism that is part of the everyday struggle for survival. As Eckersley argues, eco-centrism acknowledges that all creatures, humans included, face a daily struggle for survival, and this means that a relationship with other creatures to assure the basics of survival is often necessary and "those who adopt an ecocentric perspective will seek to choose the course that will minimize such harm and maximize the opportunity of the widest range of organisms and communities—including ourselves—to flourish in their/our own way" (57). Eco-centrism as demonstrated in the writing of Lone Wolf doesn't place humans at the center of all things; rather, it places humans in a cycle of living with all other creatures.

THE NARRAGANSETT AND WRITING THE SPIRITUAL

A common rhetorical appeal found in *The Narragansett Dawn* is an appeal to tradition, and very often tradition, and by extension the wisdom implicit in that tradition, has eco-centric dimensions. Carrying on tradition is important for the Narragansett and an often-used heading or subheading found in *The Narragansett Dawn* is, in fact, "Tradition." In the July 1935 issue, for example, one such article starts with the heading "TRADITION," followed by a subhead: "Youth Learns From Nature." This article explores spiritual connections to Narragansett Bay through the Narragansett's traditional, and very common, act of storytelling. The narrative structure of the article has a spoken-word character to it, and it begins, "Come with me for two minutes, to the banks of the Narragansett Bay, in the home of my ancestors, long before white men came" (9). Here the writer/narrator speaks directly to the reader. "Listen!" he continues as he tells the reader what he is listening to: "A father speaks to his fourteen-year old son, as they pause on the trail alone. The beauty of the sunset inspires the elder brave and the vibrations of nature help the young one understand" (9).

The story that the reader is being told is one where a father and son are retreating to the solace of the natural world for a moment of learning, allowing both father and son to reflect on the role of this particular place in their lives. As the story continues the father tells his son,

> Mishquashim, son, the hills beckon you; go forth at sunrise, when the Spirit calls. Take your bow and arrows, but do not shoot for food, until the Great Spirit commands. Look, listen and be still! The good Spirit will speak. To you will be given the mysteries of life and a symbol by which to live. Let the knowledge live in your heart, and speak only through brave deeds. (9)

In this story we again see the importance of the rising sun, marking a time when the Spirit calls. Entering the forest, the young boy is to be still and listen—an act of reflection—which will result in the answers to the mysteries of life. There is a spiritual connotation to the term "mysteries," with the Catholic religion, for instance, proclaiming the "mystery of faith" during Mass. Here the knowledge of the mysteries doesn't reside in the head, but rather in the heart, symbolic of a deeper, emotional attachment.

As the story continues, the boy is "prompted by the stirrings in his eager soul" to "go forth to seek his God alone" (10). He goes through trials and wants for warmth and food, and then "alone, amid all of nature, the little human soul appeals to the Great Spirit for a sign of understanding" (10). After a night in the forest, the story comes to a hopeful ending, again with the sun rising:

> When the first ray of light comes over the hills, he offers his prayers, as the Good Spirit has given him knowledge in his dreams. His sign is before him; and with new strength and courage, he braces himself to conquer. It may be a wolf, or deer, or it may be the beauty of a flower or cloud that first inspires him to action. That inspiration becomes his symbol and his name—like Red Cloud, Lone Wolf, or Sunflower. (10)

This story illustrates much of the Narragansett's spiritual connection to Narragansett Bay. The story begins with a call not just to the natural world but also to a particular place: "the banks of the Narragansett Bay." The boy is sent into the forest to spend a night in silent reflection, listening to the natural world around him. And it is in the warmth of the Great Spirit that he learns the knowledge of the world that is to be his. It is with the rising sun that he realizes the Great Spirit has spoken to him in his dreams. The sign that is given to the boy becomes his name, which is to say his very identity, which he draws from his location and his belonging to the ecological community around him. His sign is part of his ecological community—a wolf, or deer, or a flower. He carries that sign as part of his identity and as "a symbol by which to live."

For the Narragansett, the ecological connection to place and the creation of identity begins at birth with the very first discursive marker of identity, a name. In the September 1935 edition of *The Narragansett Dawn*, this idea is reinforced in a short article again by Lone Wolf, titled "The Baby Name." Lone Wolf explains, "When a child is born, they gave thanks to the Great Spirit in a solemn ceremony" ("Baby" 118). He further explains,

> The child was carefully watched and cared for during the year and then set out to observe nature and to learn for himself. As the glories of nature sang about him, his attention would be drawn in some direction, or to some special thing. His baby name came from this and he generally carried it until he sought his mystery to gain for himself a name. (118)

Even at the age of one, the Narragansett see it as important for a child to explore the natural world. A child, through some engagement with the natural world, takes a first identity, forming a bond to the child's ecological community. We also see from the story of the boy in the forest that at fourteen he again reflects on his place in the ecological community and is guided to a new identity. We find ecological connectedness of significant importance at two fundamental points of Narragansett identity formation: birth and the move to adulthood.

Of course, the boy's story has a pragmatic side as well. The boy, who has reached the age of fourteen, is sent into the forest to survive and find a sign. The age of fourteen likely has cultural significance. Undoubtedly he has been taught for years prior to this moment about how to survive in his ecological community. He is sent by his father, likely one of the most influential of his teachers, into the forest to reflect and connect. The connection then becomes one of spiritual relation, seeing the inspiration in the rising of the sun and the sign given to him, while also reflecting a practical knowledge of the ecosystem that allows him to survive. His short journey reaffirms a connection to, and understanding of, this particular place—the Narragansett Bay Watershed—and marks a spiritual awakening.

One of the most compelling tropes found in *The Narragansett Dawn* is the mix of traditional Narragansett spirituality and Christianity, and often there seems to be a purposeful synthesis of these two discourses. Like most New England Native peoples, the Narragansett culture was introduced to Christianity. Phillip Peckham, Church Commissioner of the Narragansett Indian Church, writes in the first issue of *The Narragansett Dawn* that it was during the New Light movement of 1741 that many Niantic tribes embraced Christianity and sought a church of their own (8). Perhaps the traditional monotheistic beliefs of the Narragansett made it easier to accept Christian beliefs. As part of the August 1935 issue of *The Narragansett Dawn*, an unidentified contributor writes,

> As we glance through the history of various nations, even when civilization was at their command, we learn how they worshipped gods of their own design, who were lifeless objects of wood or metal. But we can *proudly* claim that our forefathers worshipped the Great Spirit of the Creator of mankind. (87; emphasis in original)

Like other writers, this contributor invokes tradition to suggest a history of exception in the Narragansett and a historical alignment of traditional Narragansett belief systems with Christianity. This appeal to the historical and ancestral is again invoked in an article by Margaret Carter, titled "The Soul of the Indian," which appeared in the December 1935 issue of *The Narragansett Dawn*:

> The first missionaries in that early age of bigotry branded the Indian as a pagan and a devil worshiper and declared his life lost unless he bowed at their altars. But we of Indian blood of today are proud of the *Faith* of our ancestors in our realization that all sincere worship can have but one source and goal. (182; emphasis in original)

A more explicit example of this synthesis of traditional Narragansett beliefs and Christianity is found in the September 1935 issue, where Little

Bear makes a direct comparison between Narragansett spiritual beliefs and Christianity in an article entitled "Narragansett Fires":

> Council fires of peace, council fires of war, council fires of worship to the Great Spirit, these are not all one to the Narragansett. The first thing with them all is to signal the pleasure of the Great Spirit. They thought the Great Spirit often stood on the hills and signaled to men when the haze hung heavy. Perhaps they were not so wrong, God does speak from the hills as He spoke to Moses from a burning bush. (115)

The synthesis of traditional Narragansett spirituality and Christianity in *The Narragansett Dawn* creates a distinctly ecological variation of Christian discourse.

Consider, for example, an editorial in the July 1935 issue asks, "Do you love life, do you love these templed hills and valleys of Rhode Island, as my ancestors did? Then try the old plan—face east at sunrise and see the sunshine of God's pleasure (not yours); there you will see the DAWN of your righteous desires within your reach" (2). This editorial reveals a somewhat contrasting meaning of "We Face East." In the very first editorial from the first issue the reader is told that "We Face East" is an indicator of an eye toward the future and a new hope brought about by the New Deal, a future shaped by public policy and not spirituality. However, in the July 1935 editorial, Princess Red Wing equates the idea of turning toward the sun as it rises and feeling its warmth with an awesome spirituality. For Princess Red Wing, the sun is a reflection of "God's pleasure" shining warmly, offering a new beginning, a new day. Facing east is the "old plan" carried through generations, calling back to the wisdom of ancestry. And where does the reader stand when turning east? The reader stands in the "templed hills and valleys" of Rhode Island, with Princess Red Wing's ancestors. Here, the natural world is a temple.

Early in her article, Margret Carter establishes the connection between nature, spirituality, and monotheism. She writes,

> The Indian had many rational explanations for his religious attitude, but because he made no separation between his religious and his daily life, one does not deem his explanations always entirely satisfactory. Through nature were all the mysteries of life solved for the Indian—it was the supreme influence in his understanding of the Great Creator, called Manitou by him. (180)

Here we see Carter establishing the deep spirituality of the Indian[4] by suggesting that their religion—their spirituality—was ever present and always evident through the natural world. Carter's article offers a vision of Native Americans as deeply spiritual beings, committed to a single, monotheistic god. Moreover, Carter is concerned that people view Native American spirituality as "being smoky fire," "a totem pole," "a shrieking cry," and a "mad dance" (180). To Carter, "[m]any people believe that the Indian had no true religion"; however, as she explains, "indeed, their religion passed from the lowest stages that are known to as high a stage as that of Christianity" (180). It is not uncommon to see in the pages of *The Narragansett Dawn* the name *God* appear in proximity to the name *Great Spirit*. Often, the two are interchangeable.

Carter's next move is to distinguish between Native American spirituality and that of, although not clearly stated, Anglo-European religious practices:

> Solitary and silent amid the grandeur of the primeval forests was the worship of the Indian. The growth of a tiny seed, the abundance of food and herbs, the rain's preserving his crops,—all were the revelation of the work of an unseen being, a Creator. The sun, the stars, moon, clouds—were not these Manitou's agents to speak to man? What need had the Indian for temples and priests, since he felt that he might always meet the Great Spirit under the sun or the moonlit sky? (180)

Here, Carter removes what she sees as the Anglo-European gatekeepers to the spiritual. By removing the idea of temples and priests, she suggests that a Native American has a direct connection to the spiritual through the natural world. For Carter, it is the closeness of the Indian to the natural world that offers a spiritual connection to a supreme being.

Next, Carter again moves to a comparison between Anglo-European Christianity and Native American spirituality while giving biblical status to the oratorical tradition of Native Americans. She explains,

> What of the Bible of the Indian? The Christian and Mohammedan has his; and likewise the Red Man had his—a living book passed on from father to son, mingling history, poetry and prophecy, of precept and of folk-lore. Many of his legends furthermore, closely resemble those in Our Bible, among them, the story of the flood to purify earth—its similarity being broken only in that Noah is replaced by a virtuous Indian brave in a birch-bark canoe, and the dove, by a muskrat that dives to the bottom of the flood and brings up the earth. (180)

To conclude the paragraph, Carter asks of "[t]he Indian's story of the flood—is it not truly as feasible as ours?" (180).

Carter then in a single paragraph examines both similarity and difference around the idea of prayer. She writes,

> The one inevitable duty of the Indian was that of prayer, which was more necessary to him than to eat. Before morning light appeared entirely, he was out of his wigwam offering his prayers alone. His wife

might follow or precede him, but never did she accompany him, for each soul met the Great Spirit alone. When food was being prepared, a prayer was offered and he, who later allowed it to pass between his lips murmured, "Spirit Partake." For each thing of beauty, seen during the day, the Indian likewise offered a prayer in his silent realization of the Great Creator. The Indian had no special day for worship, since to him every day belonged to Manitou. (181)

Although Carter begins by observing a similarity between Anglo-European religion and Native American spirituality—prayer—she quickly marks the difference. First, unlike the Judeo-Christian congregational worship, Native American worship was personal and individual, meeting "the Great Spirit alone." She also suggests that the community of Anglo-European Christianity often has a mediator—clergy of some type, and she dismisses the need for a special day (perhaps making a reference to a Sabbath day). Rather, to the Native American, everyday was spiritual, and "belonged to Manitou."

Carter's comparison and contrast approach to the article accomplishes two things. First, it shows a deep and complex Native American spirituality, one intertwined in everyday life and practice. Second, despite the strong contrasts, the article shows a close connection between Native American spirituality and Anglo-European Christianity, with the key connecting component being monotheism. However, it isn't until the last paragraph of the article that Carter completes a synthesis of the two spiritual domains. She writes, "I see my forefather at sunset, his bronze body, beautifully shadowed by the sun's falling rays, in the attitude of prayer. I feel he is worshipping after all, the same Creator as I" (182). And with this final suggestion, Carter synthesizes the Creator with the Christian God, bringing together two ancient spiritual traditions to offer a unique view of God and the natural world.

CONCLUSION

During its three years, *The Narragansett Dawn* continued to serve as a counterpublic space for the continued creation and affirmation of the Narragansett's collective identity. *The Narragansett Dawn* demonstrates how the Narragansett people accepted Christianity but fused it with the traditional Narragansett belief system so to maintain the foundational eco-centric values and identity that, as told by the Narragansett, have been at the heart of the culture for generations. In the case of the Narragansett, eco-centrism, spirituality, and identity are co-constitutive. The eco-centric perspective, which is so prominent in *The Narragansett Dawn* and much of the subsequent discourses of the Narragansett, extends to all human to human interactions as well, and affirms the belief that "each human individual and each human culture is just as entitled to live and blossom as any other species, provided they do so in a way that is sensitive to the needs of other human individuals, communities, and cultures, and other life-forms generally" (Eckersley 56). In the eighty or so years since *The Narragansett Dawn* first found its way into the hands of its readership and helped create a sense of renewed solidarity, the Narragansett have fought successful battles for federal recognition and land reclamation. Still, the Narragansett continue to fight for economic sovereignty;[5] for cultural and environmental justice for themselves and for their ancestral lands; and for the well-being of vast web of ecosystems that sustain all life—human and non-human alike.

NOTES

1. The Narragansett Bay Watershed encompasses parts of southern Massachusetts and nearly the entirety of the state of Rhode Island. Narragansett Bay is the largest estuary in New England and is the defining feature of Rhode Island and has profoundly shaped the identity of the state. From the founding of the United States through the 1970s, the bay had military significance and continued regional economic significance. Narragansett Bay has also been the economic, cultural, and spiritual lifeblood of the Narragansett people for millennia.
2. Clearly the gender-biased and racially charged language used here by Princess Red Wing must be put into historical perspective, and, perhaps, is particularly problematic when framed as part of a subaltern counterpublic discourse, which I do in this essay. Despite being interpellated by these other discourses at the time, Princess Red Wing, a Native American woman, was a fierce and effective advocate for her people. Moreover, the term *Indian* is still used by the Narragansett, and can be found on the tribe's website http://www.narragansett-tribe.org. Interestingly, Narragansett cultural historians, in some cases, have resisted the term Native American.
3. Like the word *Indian*, *Tribe* is a contested word that has connotations of colonialism. However, I use the term throughout this essay as the Narragansett refer to themselves as a tribe.
4. Carter doesn't refer to the Narragansett specifically, although she does use the name Manitou, which is an Algonquin word. The Narragansett are part of the Algonquin linguistic group. Carter instead uses the more general term *Indian* to refer to Native Americans. Carter's use of the term Indian instead of Narragansett specifically is possibly indicative of an attempt to reach beyond the immediate Narragansett community to create solidarity with other Native peoples. Throughout *The Narragansett Dawn* there are common references to other Native peoples and printed correspondence from Native Americans of other tribes are frequently found in the "Letters to the Editor" or "The Narragansett Mailbox" sections.
5. In 2004, the Narragansett became embroiled in what has become known as the "Smoke Shop Raid." The Narragansett attempted to open a tax-free smoke shop on reservation land, which led to State Police intervention and a raid on the shop. A series of court cases ensued and to this day the smoke shop remains closed and feelings over the incident remain raw.

WORKS CITED

Carter, Margaret. "The Soul of the Indian." *The Narragansett Dawn* 1.8 (December 1935): 180–182. Print.

Eckersley, Robyn. *Environmentalism and Political Theory: Toward an Ecocentric Approach.* Albany: State U of New York P, 1992. Print.

Fraser, Nancy. "Rethinking the Public Sphere: A Contribution to the Critique of Actually Existing Democracy." *Social/Text* 25 (1990): 56–80. Print.

Gee, James, P. "Literacy, Discourse, and Linguistics: Introduction." *Journal of Education* 171.1 (1989): 5–17. Print.

Grinde, Donald, A. and Bruce E. Johansen. *Ecocide of Native America: Environmental Destruction of Indian Lands and Peoples.* Sante Fe, NM: Clear Light Publishers, 1995. Print.

Kern, Robert. "Ecocriticism: What Is It Good For?" *The Isle Reader: Ecocriticism, 1993–2003.* Ed. Michael P. Branch and Scott Slovic. Athens: The U of Georgia P, 2003. 258–281. Print.

Little Bear. "Narragansett Fires." *The Narragansett Dawn* 1.5 (September 1935): 9–10. Print.

Lone Wolf. "Baby Name." *The Narragansett Dawn* 1.5 (September 1935): 118. Print.

———. "Encounter with a Partridge." *The Narragansett Dawn.* 1.3 (July 1935): 18–19. Print.

———. "Little Things in Nature." *The Narragansett Dawn* 1.12 (April 1936): 285–286. Print.

———. "Trout Fishing." The Narragansett Dawn 1.1 (May 1935): 27–28. Print.

Peckham, Phillip. "The Narragansett Indian Church." *The Narragansett Dawn* 1.1 (May 1935): 8–9. Print.

Princess Red Wing. "Editorial." *The Narragansett Dawn* 1.1 (May 1935): 2. Print.

———. "Editorial." *The Narragansett Dawn* 1.3 (July 1935): 2. Print.

Robinson, Paul. "A Narragansett History from 1000 B.P. to the Present." *Enduring Traditions: The Native Peoples of New England.* Ed. Laurie Weinstein. Westport, CT: Bergin & Garvey, 1994. 79–89. Print.

Thomashow, Mitchell. *Ecological Identity: Becoming a Reflective Environmentalist.* Cambridge, MA: MIT P, 1995. Print.

Weisser, Christian, R. "Ecocomposition and the Greening of Identity." *Ecocomposition: Theoretical and Pedagogical Approaches.* Ed. Christian R. Weisser and Sidney I. Dobrin. New York: State U of New York P, 2001. 81–96. Print.

"Youth Learns from Nature." *The Narragansett Dawn* 1.3 (July 1935): 9–10. Print.

7 Conjuring the Farm
Constructing Agricultural Places in U.S. Schools

Cynthia R. Haller

Understanding the ecology of particular places can contribute to environmentally sound knowledge and practice (Billick and Price; Beatley and Manning); still, knowledge of place is largely "a construct of our symbol systems," mediated by language (Burke 4). By analyzing representations of specific places, rhetorical scholars can illuminate the social, political, and economic dynamics that construct and inform particular ecologies of place. In this chapter, I analyze how farm places are represented on *My American Farm*, an interactive website sponsored by the American Farm Bureau Foundation for Agriculture.

With less than two percent of the U.S. labor force engaged in food and fiber production, most of its population today is far removed from actual farm places. First-hand experience with rural settings, and especially farm settings, has diminished to a point that it has become increasingly difficult to even "*see* the rural at all" (Donehower, Hogg, and Schell, *Reclaiming the Rural* 3). Thus, it is doubly important to interrogate representations of farm places, to ensure that policies are informed by legitimate understandings of agriculture.

My American Farm, I argue, "conjures the farm" as an environmentally sustainable place. Further, by situating this vision of the farm within classroom ecology, the website transmits a positive *ethos* of farming to future generations of U.S. citizens, priming attitudes toward agriculture in ways that will eventually affect future agricultural and environmental policies. More generally, the website illustrates the rhetorical dynamics of ideographs in an educational forum. Thus, it can help us understand how ideologies pertaining to environmental concerns are developed, reproduced, challenged, and reconstructed in schools.

THE AGRICULTURAL LITERACY MOVEMENT: BRIDGING THE DIVIDE BETWEEN FOOD PRODUCERS AND CONSUMERS

My American Farm is designed to promote what has come to be known as "agricultural literacy." From the perspective of the Committee on

Agricultural Education in Secondary Schools, agricultural literacy is defined as an "understanding of the food and fiber system [that] includes its history and current economic, social and environmental significance to all Americans" (qtd. in Brewster 36). Although there are a variety of agriculture-related literacies, the specific term "agricultural literacy" dates back to the 1980s, when programs were developed to bring the study of agriculture into mainstream education (Brewster 36).

Agricultural literacy programs emerged partly in response to late twentieth-century changes that increasingly divided the farm sector from the consumers it served. During this time, the percentage of U.S. labor involved in farming declined steeply, for a number of reasons. Agricultural technologies were greatly increasing agricultural yields, reducing the amount of land needed for food production. In addition, the industrialization of farming had created an environment favorable to well-capitalized agricultural corporations. As high costs of farm equipment, new economies of scale, and vertical integration made it increasingly difficult for small farmers to compete, many small family farms went out of business.[1] By the 1980s, less than three percent of the labor force was directly involved with food and fiber production.[2]

Over time, the rise of agribusiness has diminished the populace's firsthand experience of farming in the U.S., making it far less likely that children growing up will spend time at the farms of extended family members such as grandparents, uncles and aunts, or cousins, let alone be brought up on farms themselves. As a result, public conceptions of agriculture have come increasingly to rely on rhetorical representations of farm places. Such representations are multiple, varying somewhat according to size, ownership, and agricultural practices of farms; and media often send conflicting messages about what a successful farm looks like and what values should govern farming practices (Ryan). Smaller family farms tend to be romanticized as pre-industrial places where people live in harmony with the land but cannot compete with corporate agriculture, or valorized as sites of "smart diversification" (Donehower, Hogg, and Schell, *Rural Literacies* 78–79). Large, corporate, industrial farms are sometimes lauded for their post–Green Revolution efficiency and ability to "feed the world"; as often, however, they are depicted as predators on small farmers, destroyers of the rural way of life, or wastelands of environmental destruction.

In the latter part of the twentieth century, negative depictions of farming began appearing with more frequency. Environmentalists in particular attacked the widespread use of chemical fertilizers, pesticides, and herbicides on farms, with agricultural corporations being portrayed as willing to sacrifice the well-being of the earth and its inhabitants for the sake of greed. Carson's *Silent Spring*, long considered one of environmentalism's seminal sacred texts, stands as perhaps the first and best embodiment of this kind of attack,[3] which, not surprisingly, precipitated defensive rhetoric from the agricultural industry and increased its perceived need to promote agricultural literacy: an "agriculturally illiterate" public, especially one disposed to mistrust the agricultural industry, could not be counted on to support agriculture-friendly policies.

Because traditional U.S. agricultural education had historically focused on preparing rural students for the occupation of farming, new avenues for supporting agricultural literacy were needed, especially those that would be of assistance in reaching urban food consumers. In the 1970s the USDA, in cooperation with the Illinois Council of Economic Education and the Illinois Women in Agriculture, sponsored a "national education project" known as *People on the Farm*, six informational booklets designed to "help students understand the nature of modern agriculture, how their food is produced, and the concept of interdependence between farmers and consumers" (*A Teacher's Guide to People on the Farm* 3). Agriculture in the Classroom (AITC), a more ambitious and long-lasting agricultural literacy program, was established in 1981 under Secretary of Agriculture John Block.

Tellingly, AITC did not originate within the division of USDA responsible for agricultural education. Rather, it was conceived and nurtured in the Office of Governmental and Public Affairs, under the leadership of the Assistant to the Secretary for Public Liaison. It can thus be identified as more a public relations than an educational program in its initial mandate. The birth of AITC can be traced to 1981, when the USDA sent letters to State Secretaries of Agriculture, asking them to establish an agricultural literacy outreach and education program in their States. Some State Secretaries delegated responsibility for AITC to the agricultural extension programs at the land-grant universities, whereas others worked with State Farm Bureaus. Even today AITC is a decentralized program, although years of national conferences and cooperation among AITC state contacts has created significant cross-fertilization among states. In 1984, Block moved AITC to the USDA's Science and Education Division, institutionalizing it by allying it more closely with traditional agricultural education programs.

In the twenty-first century, the rhetorical exigencies that gave rise to AITC have not diminished. If anything, accusations against the agricultural industry have accelerated, with charges of encouraging obesity, harming human health, and being cruel to animals added to agriculture's list of alleged sins. Industrialized agriculture-bashing publications and films are a fast-growing and popular genre (see Pollan; Estabrook; *King Corn*; and *Food, Inc.*), as are books that prescribe remedies and/or alternatives for industrialized food production and food systems (see, e.g., Hinrichs and Lyson; Hesterman). A recent *New York Times* full-page advertisement for "The Food Dialogues," an interactive public forum sponsored by the U.S. Farmers and Ranchers Alliance, reveals the extent to which the agricultural industry feels demonized. The top half of the ad page is fully consumed by a question, written in very large type: "Since When Did Agriculture Become a Dirty Word?"

As the pressure on the agricultural industry has increased, AITC programs have also grown, producing a plethora of agricultural literacy materials and resources for use in schools. Agricultural literacy programs have addressed a real need for students, most of whom have little first-hand exposure to agriculture, to learn more about the agricultural industry. As Brandt points out, however, all literacy programs are imbricated with the social, economic, and political agendas of their sponsors. Thus, as one would expect, agricultural literacy programs also send particular messages about the industry. In the next section, I discuss how *My American Farm*, an interactive website sponsored by the American Farm Bureau Foundation for Agriculture (AFBFA), works rhetorically to challenge negative environmentalist depictions of U.S. farm places. Specifically, it creates an ideographic link between farming and sustainability, generating a positive ecology of place for the American farm.

ECOLOGY OF PLACE IN MY AMERICAN FARM: INTEGRATING FARM PLACES AND SUSTAINABILITY

One of AITC's largest supporters has been the AFBFA, which provides financial and human resources to AITC programs at federal, state, and local levels. AFBFA's mission is to "build awareness, understanding and a positive public perception of agriculture through education" (AFBFA "About Us"). In a recent newsletter, Chairman Bob Stallman reminds AFBFA donors of the importance of this public perception, noting that food consumers "may not understand what it takes to provide an abundant, nutritious food supply that too many people take for granted." Alluding to negative perceptions of agriculture circulated by environmentalists and food movement critics, he notes that "farmers and ranchers often find themselves having to ask for permission to feed the world." AFBFA's vision for agricultural literacy is set forth clearly in a report on its various initiatives, including not only the *My American Farm* site but also "Ag Mags" for schoolchildren about specific products; grants for agricultural literacy programs; and the work of agricultural literacy volunteers, many of whom are Farm Bureau members. As Stallman explains, the Foundation is "engaging with kids who will one day play a vital role in shaping the agricultural systems that we'll need to maintain our food security" (1).

Recently, a revised version of *My American Farm* was presented at both the American Farm Bureau Federation's January 2011 annual meeting and the 2011 AITC conference. The primary feature of the site, its set of online educational games, was increased from five to twelve with assistance from the site's primary sponsor, Pioneer, and a secondary sponsor, John Deere. Pioneer advertises itself as the "leading source of customized solutions for farmers, livestock producers, and grain and oilseed processors." Among other endeavors, it "[provides] access to advanced plant genetics in nearly

70 countries" to help meet the goal of "increasing food production"; however, it also describes itself as a company "striving to develop sustainable agricultural systems for people everywhere" (Pioneer Hi-Bred). The site's secondary sponsor, John Deere, produces modern farming machinery.

Although information about its sponsors' high-tech activities surface within individual games, the home page of *My American Farm* is low-tech, creating a somewhat skewed picture of the modern farm. No laboratory or modern farming machine appears on the home page itself. Instead, a site visitor is greeted by blue sky, green grass, and all the accouterments of storybook farms. In the background are a red barn and silo. In the foreground are a sheep, a rooster, and a cow, familiar iconic animals of farming, surrounded by corn, cabbage, tomatoes, peppers, and grains. The background for the site's feature links has the appearance of old, weathered board; and clear, bright colors imply a cheerful, harmonious relationship among all of the depicted farm elements. Thus the home page erases industrialized aspects of the agricultural industry, invoking a romanticized vision of a pre-industrial farm.

A message to "Download your Passport to Sustainability" is also prominently displayed on the home page. This key document, a PDF file that can be folded into a passport-like structure, explicitly connects farming and sustainability:

> America's farmers and ranchers work hard to feed, clothe, and fuel the world. They take care of animals and the natural resources we have on earth, like land and water . . . so that students like you can enjoy these great things in the future.
>
> This is called "sustainability" . . . "the ability to sustain" or "to last for a very long time" . . . [Farmers] work with the soil so they can grow plants in the future! . . .
>
> Play all of the games on My America Farm to find out what you can do to support the sustainability of American Agriculture. (AFBFA, *My American Farm* "Passport to Sustainability")

The assembled Passport structure has twelve empty squares. A "passport stamp" is awarded on completion of each interactive game featured on *My American Farm*, thereby reinforcing the Passport's linkage of farming and sustainability through participatory rhetoric. In a tangrams game called "Keys to Stewardship," for example, players assemble a set of small geometric figures appropriately to "fit into" a larger geometric figure. Solving each tangram earns the player "a key and a fact," along with a round of applause, and "a picture of a stewardship practice farmers and ranchers use" then emerges within the tangram. One tangram yields a picture of *crop rotation*, which makes the soil "healthier," reduces pests, and requires "less pesticide and fertilizer." A second puzzle yields a key and fact about *drip irrigation*, which "saves energy and natural resources like water." A picture of a drip irrigation nozzle appears at the base of a plant with a message that "Farmers and ranchers are America's #1 stewards of the land." Two other tangrams feature rotational grazing, which "allows natural grasses and other plants to grow stronger and healthier while livestock graze in other pastures," and wildlife food plots, reserved farmlands that serve as space for wildlife habitation. Once the player has finished four tangrams successfully, a printable page pops up with "Congratulations" in large type, followed by a note about what the player has won. This printable page contains a passport stamp (along with an enlarged picture of the stamp) depicting "contour farming," a planting method designed to "reduce erosion and conserve rainwater." It also includes a short passage further elaborating the stewardship theme of the game; instructions for a supplementary hands-on activity that reinforces the theme of the game; and some "cool facts" or further information about the theme (see Figure 7.1).

My American Farm 's repeated references to sustainability both draw upon and underscore the emerging rhetorical power of the sustainability paradigm in environmental discourse. In the 1990s, Killingsworth and Palmer characterized environmental policy arguments as highly polarized, driven by Barry Commoner's "environmental dilemma" between material progress and environmental protection (*Ecospeak*). In the twenty-first

CONGRATULATIONS

YOU WON a FIELD DESIGNED USING CONTOUR FARMING. CONTOUR FARMING IS THE PRACTICE OF PLANTING ALONG THE CONTOUR, OR "SHAPE" OF THE LAND TO REDUCE EROSION AND CONSERVE RAINWATER.

A SLICE OF SOIL!

Soil is one of our most important natural resources on the earth's surface. Complete this activity to learn just how little soil we have to grow our food.

You'll need an apple, a paring knife, and the help of an adult.

- Cut an apple into four equal parts. Three parts represent the oceans of the world. The fourth part represents the land area.
- Cut the land section in half lengthwise. Now you have two 1/8 pieces. One section represents land such as deserts, swamps, Antarctic, Arctic, and mountain regions. The other 1/8 section represents land where man can live and may or may not be able to grow food.
- Slice this 1/8 section crosswise into four equal parts. Three of these 1/32 sections represent the areas of the world that are too rocky, too wet, too hot, or where soils are too poor to grow food. Plus, we can't grow food on some land because cities and other man-made structures are built on it.

- Carefully peel the last 1/32 section. The peel on this small piece represents the amount of soil on which we have to grow food. This amount of soil will never get any bigger.
- Consider this: With so little soil and so many people on the earth, how are we able to grow enough food to feed everybody? After completing this activity, do you feel it is important to conserve the soil we use to grow our food or not?

Farmers throughout the United States take great pride in caring for our land. Farmers know the land better than anyone else. They know what the land needs to be healthy, so that they can grow nutritious crops for you and me. A farmer's land is also valuable to them, so it is in their best interest to keep it in good condition. According to the National Cattlemen's Beef Association, "For generations, our families have lead conservation efforts across the United States. We know environmental stewardship and good business go hand-in-hand."

CUT OUT THIS stamp and attach it to your passport to sustainability. collect all 12.

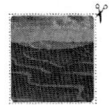

OTHER COOL FACTS ABOUT STEWARDSHIP

- The Environmental Quality Incentives Program, also called EQIP, gives farmers and ranchers help to protect soil and water quality.

- Farmers and ranchers who have livestock practice good stewardship by managing grazing. Animals are allowed to graze (eat the grass) on certain fields, sometimes in a rotation, to help protect the soil and water in that area.

Figure 7.1   "Keys to Stewardship" reward map. (Source: AFBFA *My American Farm.*)

century, however, the sustainability paradigm has largely supplanted the environmental dilemma as an ideological touchstone for environmental policy arguments. Killingsworth and Palmer themselves predicted that, in a culture centered on environmental values, the dialectical struggle of developmentalists and environmentalists would be transcended by "a new discourse [that would] . . . emerge as the metanarrative or mythology by which the culture carries its values across generations" (265). Although the world has yet to attain Killingsworth and Palmer's hoped-for level of eco-logical consciousness, the paradigm of sustainability has, to a great extent, fulfilled their prediction of the emergence of a new mythology. Words such as "conservation," "stewardship," "sustainability," "green," and others have become a set of God-terms against which both environmental policy and economic development activities are now measured. Sustainability dis-course, a relatively new phenomenon when *Ecospeak* was published, is now the preferred language of both environmentalists and developmentalists.[4]

My American Farm capitalizes on the power of these new God-terms through an ideographic dynamic. Ideographs, as McGee defines them, are essentially a "vocabulary of concepts that function as guides, warrants, reasons, or excuses for behavior and belief" (6). More "pregnant" than propositions, they work as "God" or "Ultimate" terms, tacit beliefs that dispel the need for argument because they contain whole arguments within themselves. Ideographs are "one-term sums of an orientation," "*logical* commitments," taken as true without requiring the support of argument (7). Deeply embedded in a community's psyche and passed down over time, they persuade effectively because they persuade unconsciously.

In his discussion of ideographs, McGee focuses on obviously ideologi-cal concepts like liberty, equality, and so on. Agriculture, however, is an ideographic term as well, at least in the U.S. Agriculture's ties to morality and civics go back at least to the early republic, when Jefferson wrote of the inherent virtuousness of farmers; and farming has been ideologically inter-twined with politics and economics ever since. It has always held a moral dimension, although whether it is defined as morally good or morally bad shifts according to time and place (Hagenstein, Gregg, and Donahue 4–6). After the percentage of the labor force involved in farming slid to under three percent in the late twentieth century, for instance, "back to the land" movements of the 1970s promoted a return to agriculture as a return to virtue. In the present time, the connection of agriculture to virtue has been more schizophrenic. Organic farming, along with consumption of health-ful, locally produced foods, has come to be associated with moral virtue, even as conventional, industrialized agriculture has come to be associated with moral turpitude (369–376).

Ideographs, according to McGee, are structured both diachronically and synchronically. The diachronic (vertical) structure of a community's ideo-graphs (the sum of its contextual meanings as accrued over time) is primar-ily preserved and passed down in school. Indeed, McGee cites "grammar school history" as the "most truly influential manifestation" and record of the vertical structure of ideographs, as exemplified by the persistence of historian William Wirt's misattribution of the phrase "give me liberty, or give me death" to Patrick Henry. According to McGee, the story, although inaccurate, endures because it is less important as truth value than as an "ideographic touchstone" for liberty, influencing how students will later understand and make "judgments of public motives and their own civic duty" (11).

If schools maintain the diachronic structure of ideographs, rhetoric manages their synchronic structure. At any given time, a society's ideographs are arranged into ideographic clusters, or sets of ideographs connected to and aligned with one another. Ideographs within this horizontal, synchronic structure take on the character of "forces," operating in often confrontational ways and always "com[ing] to mean" in the present time. Rhetoric works to alter and reshape ideographic clusters, which can "[be] restructured, perhaps broken, in the context of a particular controversy" (12).

My American Farm's connection of farming and sustainability, however, suggests that schools do not simply preserve and transmit the diachronic structure of ideographs; they can also be sites for synchronic rhetorical intervention in ideographic clusters. The Passport to Sustainability, for instance, functions rhetorically by breaking a negative ideographic connection (agriculture=environmental harm) and establishing a positive one (agriculture=sustainability). Further, it accomplishes this ideographic rearrangement not only through words, but also through performative rhetoric. As a player completes each game successfully, he/she is awarded a passport stamp that can be glued onto the sustainability passport. Gameplayers themselves thus construct the farm-sustainability linkage through their own actions, eventually completing a passport that conjures a sustainable virtual farm.

In terms of McGee's example of the Patrick Henry misattribution, the Passport to Sustainability might be compared to a staged miniplay where children enact the Patrick Henry drama rather than simply read about it. Through enactment, body and mind together reproduce the ideograph. Similarly, completion of the Passport to Sustainability reproduces the existing structure of the sustainability paradigm. However, it also enacts a new ideographic cluster, in which agriculture is unilaterally yoked to sustainability.

The twelve games on *My American Farm*, together with the Passport to Sustainability, thus construct a particular ecology of place for the American farm. They suggest that it is characterized by sustainable practices and stewardship. An advantage of this vision is that it counteracts the unhelpful, unilateral villainization of the agricultural industry that has become common practice among today's critics of U.S. food systems. On the other hand, *My American Farm* falls prey to unilateralism as well. Specific agricultural places employ different practices and yield differing environmental impacts, but such differences are downplayed on the site. Instead, environmentally sustainable practice is depicted as an integral part of "The Farm's" identity.

ECOLOGY OF PLACE IN EDUCATION: BRINGING AGRICULTURE INTO THE CLASSROOM

As described in the previous section, *My American Farm* rhetorically constructs a particular ecology of place for the U.S. farm. The power and reach of its idealized farm, however, is amplified by the site's rhetorical reconstruction of a different place: the classroom. Schools, as McGee suggests, are nurseries for the propagation of human values. Seedbeds of ideologies, they implicitly shape the course and character of the future. Accordingly, the environmental stakes in education are at least as high as those in policymaking forums. At stake in the classroom are not current but future policy decisions, which will be determined in part by the ideological stances instilled in children toward environmental issues. It's important, then, to understand not only how *My American Farm* conjures the farm, but also how it makes space for that farm in the ecology of the classroom.

Agricultural education has not traditionally been part of mainstream elementary and secondary curricula in the U.S., except in rural areas and/ or as a vocational curriculum. AITC proponents, then, to be successful in promoting agricultural literacy, must create a space for the study of agriculture that fits appropriately within all educational ecosystems. *My American Farm* accomplishes this goal by appealing to disciplinary divisions and curriculum standards, the special topics of argument that currently govern the organization and delivery of curricula in U.S. schools. [5]

In the "Educators Resources" section of *My American Farm*, teachers can access the site's content through two links: (1) Curriculum Area and (2) Ag Topics. Two of the categories on the "Ag Topics" link—"Farmers Care for Animals" and "Farmers Steward the Land"—serve to reinforce *My American Farm*'s redefinition of the ecology of the farm, showcasing the farm as friendly to both the environment and animals. The "Curriculum Area" link, however, accomplishes a different form of rhetorical work. It cross-references the site's content material with traditional elementary and secondary curricular divisions: math, science, social studies, English language arts, and health.[6] *My American Farm* thus identifies agricultural literacy as squarely within the existing disciplinary ecology of schools.

Also in the "Educators Resources" section, the "Educational Standards Crosswalk" cross-references the site's various activities with specific curriculum standards, showing teachers how they can use agricultural topics to meet existing curricular goals. In U.S. elementary and secondary schools, curriculum standards define the pedagogical ecology of the classroom; it is expected that teachers use classroom lessons and activities to assist students in reaching these standards.⁷ By aligning its activities with curriculum standards, *My American Farm* provides teachers with a recognizable, accepted pedagogical rationale for agricultural literacy. The curricular matrix presents agricultural literacy as a means for meeting already existing curricular goals rather than as a separate curricular area—an easier ecological adjustment to propose than creating an entirely new space within classroom ecology.

CONCLUSION

The foregoing analysis illustrates how an ecology of place, in the case of the American farm, can be constructed ideographically. Most children today (as well as their parents) have little personal experience of farming, making their perceptions highly dependent on how agricultural venues are presented in the media. As one representation of farm places, *My American Farm* exemplifies how ideographs can be used to build positive *ethos* for a place, as well as how ideographic clusters can be rhetorically (re) constructed to support that *ethos*. *My American Farm* effectively counters negative stereotypes of industrial agriculture as a cabal of corporations that ruin the environment and grow corrupted food and fiber. By "conjuring the farm" as a place that incorporates stewardship of the land and sustainable practices, the website creates a positive perception of farming. In so doing, however, it tends to erase the many differences among agricultural methods and practices across the U.S. Thus local, material conditions of specific agricultural ventures, which are necessary to determine sound environmental policy in particular situations, are glossed over. Ultimately, such glossing of difference can hinder the optimal application of ecology of place to agricultural planning and policy.

As my analysis of the "Educator Resources" section of *My American Farm* suggests, rhetoric can also influence and alter perceptions of how a specific ecology of place is related to broader ecosystem hierarchies and networks. By recategorizing agricultural content knowledge according to already accepted curriculum divisions, the "Educator Resources" present farm places as nested squarely within the ecology of education. By aligning proposed classroom activities and resources with specific curriculum standards, the site makes it easy for teachers to see where farming fits into the school sanctioned school goals. Given the challenges of standards-based teaching, the strategy is highly effective. Teachers, already faced with a crowded school day, do not need to "make room" for agricultural literacy; they can simply use agricultural content to help students achieve the designated standards.

Aside from illustrating how ecology of place can be constructed ideographically, this study also speaks to twenty-first century shifts in environmental rhetoric strategies. In *Ecospeak*, Killingsworth and Palmer describe a great divide between developmentalists and environmentalists, one that hinders the development of sound environmental policies. The rise of sustainability discourse, however, is changing the character of environmental debate in today's "green" era. Instead of taking polarized, oppositional sides, both developmentalists and environmentalists search for arguments within a middle ground of shared terms. God-terms of sustainability discourse—conservation, stewardship, environmental protection, and so on—are frequently invoked to support the *ethos* of both environmentalists and developmentalists, sometimes making it difficult for the public to discern differences in their policy stances.

In some ways, this paradigm of sustainability has been productive for environmentally sound policy setting: its terms enable environmentalists and developmentalists to argue about their concerns in a shared language. In the best of circumstances, the language provides common ground on which to successfully negotiate difference in particular policy-setting situations (Norton). Certainly, the pitfalls of using incommensurable discourses for debating environmental policy have been well illustrated. Killingsworth and Palmer describe how a discourse of instrumental rationality marginalized the voices of small farmers, Native Americans, and even a "big-time rancher/banker" in the case of the *East Socorro Grazing Environmental Statement* (*Ecospeak* 188). Peterson analyzes how the incommensurable discourses of government, Agriculture Canada, and aboriginals derailed wood bison management policy in Canada, preventing the emergence of "collaborative definitions of sustainable development" during the hearings (115).

The sustainability paradigm, however, is no panacea for environmental policy. Appeals to sustainability can sometimes obscure very real differences between policy positions competing for the validation of public audiences. Logical appeals to stewardship and sustainability, for instance, can be used to argue for conflicting positions in a discussion of whether and how to regulate agricultural use of fertilizers. On the one hand, fertilizers increase yield and thus minimize the need to develop land, a natural resource, for agricultural purposes. On the other, runoff from fertilizer use endangers water resources. With logical appeals to sustainability capable of supporting very different policy goals, the importance of appeals to *ethos* increases. The public's identification with a particular side of a controversy may well be determined by which side successfully creates an *ethos* of acting in the interests of sustainability.

Finally, with respect to rhetorical theory, the ideographic strategies of *My American Farm* complicate McGee's depiction of schools as sites for preserving and passing down the diachronic structure of ideographs. Schools are also sites of rhetorical intervention in the synchronic structure of ideographs, places where existing ideographic clusters can be challenged, broken up, and rearranged as well as reproduced. Ideographs, apparently, "come to mean" (McGee 12) according to the specific ways they are employed in school forums. Agricultural organizations understand the rhetorical stakes involved in classrooms, and are using that understanding to reshape children's perceptions of agriculture. Rhetorical scholars should follow their lead by investigating more generally how ideographic clusters are invented, maintained, and changed in school settings. The rhetoric of education is literally the rhetoric of the future.

NOTES

1. For more detailed analysis of the factors contributing to the rise of corporate and global industrial agriculture and its impact on family farmers, see Schell, "The Rhetorics of the Farm Crisis," 84–92 in Donehower, Hogg, and Schell, *Rural Literacies*.
2. The decline in numbers has continued into the present time. Today, farmers compose less than two percent of the workforce.
3. Killingsworth and Palmer (*Ecospeak*) have argued that *Silent Spring*, which portrays developmentalists and agriculturalists as enemies of environmentalism, is a good example of the divisive environmental rhetoric of "praise and blame" (76). See also Killingsworth and Palmer, "Millennial Ecology" and Slovic for detailed analyses of Carson's rhetoric.
4. As an example of the pervasive quality of the sustainability paradigm, we need look no further than scholarship in rhetoric and composition, where the concept of eco-composition has informed our theories of the writing (see Owens; Dobrin and Weisser).
5. For a fuller discussion of special topics, see Leff, and Miller and Selzer.
6. The so-called enrichment areas (e.g., music, art, physical education) are not on the list. These areas are (unfortunately) considered less essential to school curriculum: they are often the first to be cut when school budgets come under the axe.
7. At the time of this writing, national standards, known familiarly as the "Common Core," are in place for both language arts and mathematics. In other disciplines, such as science and social studies, states still define their own standards.

WORKS CITED

American Farm Bureau Foundation for Agriculture. "About Us." *My American Farm*, n.d. Web. 17 Dec. 2011.
———. *My American Farm*. N.d. Web. 17 Dec. 2011.
Beatley, Timothy, and Kristy Manning. *The Ecology of Place: Planning for Environment, Economy, and Community*. Washington, DC: Island P, 1997. Print.
Billick, Ian, and Mary V. Price. *The Ecology of Place: Contributions of Place-Based Research to Ecological Understanding*. Chicago: U of Chicago P, 2010. Print.
Block, John. "Secretary's Memorandum 1020–14." 5 April 1984. TS. USDA History Collection, Series 3, Box 2, Folder Secretary's Memos, 1921–1984. National Agriculture Library, Beltsville, MD. Print.
Brandt, Deborah. *Literacy in American Lives*. Cambridge: Cambridge UP, 2011. Print.

Brewster, Cori. "Toward a Critical Agricultural Literacy." *Reclaiming the Rural: Essays on Literacy, Rhetoric, and Pedagogy.* Ed. Kim Donehower, Charlotte Hogg, and Eileen E. Schell. Carbondale: Southern Illinois UP, 2012. 34–51. Print.

Burke, Kenneth. *Language as Symbolic Action.* Berkeley: U of California P, 1966. Print.

Carson, Rachel. *Silent Spring.* Boston: Houghton Mifflin, 1962. Print.

Dobrin, Sidney I., and Christian R. Weisser. *Natural Discourse: Toward Ecocomposition.* Albany: State U of New York P, 2002.

Donehower, Kim, Hogg, Charlotte, and Eileen E. Schell, eds. *Reclaiming the Rural: Essays on Literacy, Rhetoric, and Pedagogy.* Carbondale: Southern Illinois UP, 2012. Print.

———. *Rural Literacies.* Carbondale: Southern Illinois UP, 2007. Print.

Estabrook, Barry. *Tomatoland.* Kansas City, MO: Andrews McMeel, 2011. Print.

Food, Inc. Dir. Robert Kenner. Magnolia, 2008.

Hagenstein, Edwin C., Sara M. Gregg, and Brian Donahue, eds. *American Georgics: Writings on Farming, Culture, and Land.* New Haven, CT: Yale UP, 2011. Print.

Hesterman, Oran B. *Fair Food: Growing a Healthy, Sustainable Food System for All.* New York: PublicAffairs, 2001. Print.

Hinrichs, C. Clare, and Thomas A. Lyson. *Remaking the North American Food System: Strategies for Sustainability (Our Sustainable Future).* Lincoln: U of Nebraska P, 2009. Print.

Killingsworth, M. Jimmie, and Jacqueline S. Palmer. *Ecospeak: Rhetoric and Environmental Politics in America.* Carbondale: Southern Illinois UP, 1992. Print.

———. "Millennial Ecology: The Apocalyptic Narrative from *Silent Spring* to *Global Warming.*" *Green Culture: Environmental Rhetoric in Contemporary America.* Ed. Carl G. Herndl and Stuart C. Brown. Madison: U of Wisconsin P, 1996. 21–45. Print.

King Corn. Dir. Aaron Woolf. Mosaic, 2006. Film.

Leff, Michael C. "The Topics of Argumentative Invention in Latin Rhetorical Theory from Cicero to Boethius." *Rhetorica* 1 (Spring 1983): 23–44. Print.

"Maryland Teachers Must Promote Environmental Literacy." Narr. Larry Abramson. *Morning Edition.* Natl. Public Radio. WNYC, New York, 15 Sept. 2011. Radio.

McGee, Michael Calvin. "The 'Ideograph': A Link between Rhetoric and Ideology." *Quarterly Journal of Speech* 66 (1980): 1–16. Print.

Miller, Carolyn, and Jack Selzer. "Special Topics of Argument in Engineering Reports." *Writing in Nonacademic Settings.* Ed. Lee Odell and Dixie Goswami. New York: Guilford, 1985. 309–341. Print.

Norton, Bryan G. *Sustainability: A Philosophy of Adaptive Ecosystem Management.* Chicago: U of Chicago P, 2005. Print.

Owens, Derek. *Composition and Sustainability: Teaching for a Threatened Generation.* Refiguring English Studies. Urbana, IL: National Council of Teachers of English, 2001. Print.

Peterson, Tarla Rai. *Sharing the Earth: The Discourse of Sustainable Development.* U of South Carolina P, 1997. Print.

Pioneer Hi-Bred. *Pioneer Hi-Bred Introduces 33 New Soybean Products for 2012.* Iowa: Pioneer Hi-Bred, 8 Nov. 2011. Web. 3 Jan. 2012.

Pollan, Michael. *The Omnivore's Dilemma: A Natural History of Four Meals.* New York: Penguin, 2006. Print.

Ryan, Cynthia. "Get More from Your Life on the Land." *Reclaiming the Rural: Essays on Literacy, Rhetoric, and Pedagogy.* Ed. Kim Donehower, Charlotte

Hogg, and Eileen E. Schell. Carbondale: Southern Illinois UP, 2012. 52–71. Print.

Slovic, Scott. "Epistemology and Politics in American Nature Writing. Embedded Rhetoric and Discrete Rhetoric." *Green Culture: Environmental Rhetoric in Contemporary America.* Ed. Carl G. Herndl and Stuart C. Brown. Madison: U of Wisconsin P, 1996. 82–110. Print.

Stallman, Bob. "Your Support Produces Engagement and Collaboration." Washington, DC: American Farm Bureau Foundation for Agriculture, n.d. Print.

A Teacher's Guide to People on the Farm. 14 Sept. 1981. TS. USDA History Collection, Series 1.5, Box 29, Folder X. National Agriculture Library, Beltsville, MD. Print.

U.S. Farmers and Ranchers Alliance. "The Food Dialogues." Advertisement. *New York Times,* 19 Sept. 2011, A5. Print.

8 Digital Cities
Rhetorics of Place in Environmental Video Games

Michael Springer and Peter N. Goggin

Monkey Island; Washington, DC in 2277; Jackson Square, New Orleans; Tatooine—other than puzzle genres (scrabble, crosswords, chess, etc.), most video games are dependent for story, context, action, and adventure on some design of constructed place. That is, some element of a virtual or representational geographic or conceptual location. It may be the highly bounded imaginary or representational location of an auto racing arena (*Grand Prix*; *F1 Race Stars*; *Super Mario Kart*) or sports game (*Mike Tyson's Punch-Out!*; *Space Jam*; *NBA 2K13*), carefully reconstructed representations of real physical places such as Jackson Square (*Gabriel Knight: Sins of the Fathers*) or Florence during the Renaissance (*Assassin's Creed II*); reimagined physical places that are altered by time and circumstance such as Washington, DC in 2277 (*Fallout 3*); imagined places in an otherwise real geographic locations such as Monkey Island in the Caribbean (*The Secret of Monkey Island*) or places that exist only in the imaginations of the game writers and designers such as Tatooine (*Star Wars: Knights of The Old Republic*).

As with many other media forms, notions of place and space are as fluid as our perceptions and definitions will allow. Virtual or digital environments, like photographs, scripted texts, broadcast media, iconographs, sign and spoken language, and so forth, are not places in the geophysical material sense but are places, or representational of places, because we, humans, make them so. Accommodating this view allows us to move beyond the philosophical conundrum of the old "if a tree falls" saw and consider the relevance of rhetorical constructions of place as they inform virtual environments or places, and how place-based rhetorics may inform the design of digital gaming toward pedagogies of sustainability.

Video games have a lot to teach us, and we can learn a lot from video games. So go the claims of some very prominent educators, scholars, and researchers in the rhetorics and literacies of video games and gaming theory (Bogost; Gee; Murray; Hayes and Duncan; McGonigal; Aarseth). Until relatively recently the academic significance of gaming has largely been the province of engineering design, and mathematical/computational theory and philosophy (and, of course, on the commercial side, marketing and entertainment). Over the past decade, however, video gaming is starting to be recognized more broadly by academics and the public as a powerful educational medium. Events such as the annual Games for Change festival, where designers compete over who has designed the best game with a social message, reflect this growing awareness of the potential games hold to inform and influence our thinking. Meanwhile, the annual Games for Health conference focuses on how video games can be utilized for improving health and healthcare. Within academia, journals such as *Game Studies*, *Games and Culture*, and *ELUDAMOS* are chronicling research on the socio-cultural dimensions of video games. Academic centers such as Arizona State University's Center for Games & Impact, UC Irvine's Center for Computer Games & Virtual Worlds, and a variety of others serve as loci for ongoing research into how video games affect culture and education. As a part of this larger movement, more games are being released that purport to teach players about environmental sustainability, including *World Without Oil*, *PowerUp*, *City Rain*, and *MiniMonos*.

However, one of the crucial challenges for gaming design when it comes to teaching about environmental concerns and sustainability is the virtual representation of natural systems and place. Synthetic places are less problematic. Gaming locales that are based on actual or fictional man-made places and spaces, museums, arenas, cities, space stations, parks, gardens, and so forth can be impressively detailed and realistic. The *Gabriel Knight* and *Sherlock Holmes* series, for instance, provide almost broadcast video quality of such places as Neuschwanstein Castle in Bavaria (*Gabriel Knight II: The Beast Within*) and the British Museum and National Gallery (*Sherlock Holmes: Nemesis*). *Assassin's Creed II* gives players the chance to stealthily sneak through a partially built Basilica di Santa Maria del Fiore, along with other buildings from Florence in the late fifteenth century. Players of games such as these, even though the primary goal is to solve a mystery, will find themselves on virtual tours of real-world places. Art, history, architecture, literature, and sciences are woven into the virtual experience of the game play.

Extended external, natural, and man-made places fare less well. The ecological diversity of cities and natural systems, even those that are designated, conserved, and preserved by humans such as national parks and wilderness areas cannot (for now) be fully represented by gaming design. The result is that place in these cases is ecologically limited to very few generic species of flora and fauna, and cityscapes and landscapes are sterile and repetitive. Like the scrolling background for animated cartoons, the natural environmental places and spaces for games typically serve as a backdrop for game play or as obstacles that must be avoided, "beaten," or destroyed. If the concept of "globalization," that is, the flattening or "Walmarting" of the world, can truly be said to apply anywhere, it is in the virtual ecology of video games where biodiversity, cultural diversity, and geo-physical diversity are extremely limited. The key to immersive game play in these contexts is to ignore ecological representation and not to look too closely at one's surroundings.

Coupled with the challenge of creating environmental place-reality in virtual space (which most game designs necessarily avoid) is the challenge of designing games that are intended to actually provide pedagogical value. Many games such as those featured at the Games for Change festival that are developed with teaching in mind are limited by prioritizing the wrong audience. Often these games designed for educational purposes are created to be played all the way through in one class session, giving players a fraction of the immersive experience they might get playing a commercially released console game. Another game element that can interfere with a sense of immersion is when the educational elements of the game are too glaringly obvious. "Educational" is synonymous with "boring" for many young gamers, and most of us can understand why when we remember how much less we enjoyed reading novels as part of high school classes than when we self-selected them because we *wanted* to read them.

If the medium of video games is to be used as a successful tool for teaching about environmental issues, these games must be designed to create virtual places that although not necessarily environmentally "real" in terms of bio- and geo-diversity are still highly interactive and challenging. It must feel like a real video game, not an educational one. Instead of being designed with teachers and classrooms in mind, which is often the case in order to legitimize them for school-based learning purposes, these games should cater to the players.

Notably it has been corporations rather than educational institutions that seem to best understand this concept. As Gee observes, capitalists—"the makers of video games"—are not limited by polemic education politics that distinguish between "overt information and immersion in practice." Gee suggests that gaming companies inherently understand that these learning approaches inform each other. "Good video games incorporate good learning principles, because otherwise there would be no video games, because too few people would have purchased them" (113–114). IBM, for instance, attempts to merge the notions of player-centered interests and challenging learning practices as central to game-play with its sustainability education game, *PowerUp.*[1] As Steven Walvig, director of education at the Bakken Museum in Minneapolis states about this, and other would-be environmental education games:

> To be successful, an educational game needs to have a social piece . . . You get interested and you talk about it. We want people to take that experience home and talk about it at the dinner table or at school the next day. It's not just about spreading the word. When kids start explaining their experience, that's when they really get it. The learning is more effective when it's fun and social. (Libby)

Yet, the challenge remains. In addition to *PowerUp* and other sustainability-themed educational games such as *City Rain*, *Plan It Green*, and *World Without Oil*, several recent games have focused specifically on the problem of energy consumption and the difficulties in providing enough energy for cities. In the remainder of this chapter we discuss how two such games, *Energy City* and *Energyville*,[2] create digital places where important learning can occur: virtual spaces where gamers are immersed in real-world problems. Both games have clear rhetorical arguments and attempt to communicate complex ideas to their audiences. But at the same time, design decisions limit both the rhetorical effectiveness and the playability of these games.

A TALE OF TWO CITIES: ENERGYVILLE AND ENERGY CITY

Two key principles, Gee's notions of active critical learning/meta-critical thinking (Gee offers thirty-six principles in all), and virtual-to-real-world-transference (Bogost's notion of procedural rhetoric), provide a robust theoretical frame for analyzing the effectiveness of these games for learning about environmental sustainability. This frame allows us to ask questions about the digital representation of cities, how these digital cities function, how they are affected by environmental threats, and what virtual-to-real-world-transference is encouraged by in-game feedback.

Both *Energyville* and *Energy City* are management simulations, a genre made popular by titles like *Sid Meier's Civilization*, *SimCity*, and *Age of Empires*. Games within this genre allow players to make decisions about construction and economics for a city or nation, working to keep the population happy while also accomplishing other game goals. Some titles involve making decisions about war and conquest. Others set a particular technological achievement as the final goal. *Energy City* and *Energyville* are simple versions of this game type. Players are in charge of the energy portfolio for a city, making decisions about how to generate the energy required to keep the city running while also taking the environmental impact of these decisions into account. Both games put the player in a somewhat generic urban area, allowing players to draw connections between the issues being faced in their simplified game world and the issues faced by real cities.

Energy City is a game made available online by the JASON Project, a nonprofit organization that designs authentic experiences where individuals can learn more about science and our world. Their projects extend to partnerships with National Geographic and the Sea Research Foundation, and go beyond the field of video games into physical places, including Mystic Aquarium, one of the world's largest aquariums ("About JASON"). A broad variety of games are available on their website, many of them entirely free. *Energy City* allows players to take up the role of a city's energy manager, a multifaceted position that requires the careful juggling of environmental concerns with energy needs, financial restrictions, and public opinion.

Players are challenged to power a city without harming the city's air quality or doing unsustainable damage to global environmental health. While accomplishing this, players must also keep track of their energy spending and address the myriad requests from interest groups. Without public support for the player's actions, the public will be uncooperative, and this will directly affect the funding for new energy initiatives. Thus, ignoring the political aspect of energy policy is likely to create an irritated citizenry who make sustainability more difficult. To win the game, players must meet the energy needs of the community while successfully keeping pollution and spending to sustainable levels.

At the beginning of the game, players choose a particular city they wish to play, each of which has a different difficulty level and different physical appearance. Several cities are small and in grassy areas, while others are located in deserts. Implicit in this choice is the fact that some cities are harder to power than others; limited access to resources has a direct impact on how difficult it is to generate energy. Some cities are capable of building wind turbines that produce a significant contribution to the energy portfolio. In other cities, where wind is apparently not as plentiful, the same turbines generate significantly less power. While some cities are located in areas where hydro-dams could be constructed, others do not have this option. Each city, then, requires the player to strategize based on resource availability.

The least expensive option for powering cities is to continue using the options already in place at the game's beginning: coal and oil. These plants can generate a lot of energy quite easily, allowing the player to complete each turn quickly without spending much money. However, this is not a sufficient long-term strategy for winning the game, because global environmental health and air quality are both reduced when these plants are operated. The beginning of the game, then, will show players a city that cannot help but get a little less healthy with the current resources. However, if players choose to spend the substantial money required, they can begin the construction of plants that allow for biofuel, geothermal, tidal, or wind energies. It takes several years—or turns—for each of these plants to become operational, making them long-term investments. However, once more energy can be produced by less harmful methods, the environmental condition begins immediately improving.

Conservation is another method for meeting some of the city's energy demands. Commercial and residential planning are two options that reduce this total energy need. Transportation planning, bike paths, and rooftop gardens can all reduce the total amount of energy required by the city. However, with conservation strategies, players also need to make an initial investment that will only pay off gradually. Again, players are encouraged to recognize a correlation between initial investment and long-term, gradual improvements.

Through careful decisions and strategy, it's possible to eliminate the use of nonrenewable resources and get by with a combination of conservation and renewable energy. This leads to the environment improving dramatically over the remainder of the game. At the end of each turn, which represents one year, charts show how successful the player has been up to this point. One graph illustrates how much energy is provided by each resource type. Another graph illustrates how the air quality has been in the city throughout the game, and whether it is improving or getting worse. Another graph shows the city's global impact on health as well as how successful the player has been in conserving money.

Meanwhile, several interest groups ask for favors throughout the game, offering incentives for working with them. Middle school children are one of these interest groups and often have requests based on things they've seen in nearby cities. On occasion, they'll ask the player not to build a tidal power plant because it'll make it harder to go fishing. Sometimes, they'll want garden rooftops, which have a variety of ecological benefits. The children want them simply because they look cool.

The other interest groups often make suggestions as seemingly arbitrary as these. The local businesses will sometimes insist the player spend a lot of money that turn because "[y]ou have to spend money to make money." In other turns, they will complain about reckless spending and ask that expenditures be kept below a certain dollar amount. Regardless of how impulsive their requests seem, there is significant pressure to go along with them because the benefits of doing so could assist or impede the player's success. The game is thus complicated by the desires of interest groups, and sometimes it seems that the most effective way of winning the game is to be always doing what the citizens want. In order to win the game, players of *Energy City* must keep all three barometers—environmental health, air quality, and budget—above zero. If this is failed at any point, the game abruptly ends, suggesting some options that may make them more successful the next time they play.

Chevron's *Energyville* asks players to achieve the same primary goal: generate enough power to keep a city functioning without going over budget or doing an excessive amount of damage to the environment. In some other significant ways, the games are nearly identical: the player has barometers that show environmental impact and financial impact, and a variety of energy types can be used to reach the necessary amount of power. But rhetorically, the games are quite different. *Energyville* allows players to make a small number of structural changes to their cities, and requires a certain proportion of energy to be generated by petroleum. Any attempt to avoid using petroleum for an entire turn will be met with a warning: "YOUR CITY NEEDS PETROLEUM. Although alternative fuels can reduce the need for petroleum, airplanes and a significant portion of ground vehicles will continue to rely on petroleum for fuel." At this point, players cannot advance any further without using petroleum to fulfill the city's remaining energy needs.

Energyville does not have the option of creating initiatives that will cut down on the amount of energy required. The issue is instead one of merely meeting the city's demands, not reducing them. Players aren't dependent on public opinion, and are allowed to make decisions without considering anything beyond the impact meters. This makes the game a less immersive experience than *Energy City* where players are required to remain a respected decision maker on top of providing energy and reducing environmental impact.

A pie chart shows the players how much of their energy needs are being met by each type of resource, and a meter beside this pie chart shows how much energy is still needed. Small meters along the right side of the screen show the player's environmental impact, financial impact, and security impact. Energy security refers to how diverse the energy resources available to the city are, as well as how reliable these resources are determined to be by the game's designers. The security of these resources comes into play when time passes within the game, which leads to randomly selected events impacting the availability and cost of specific resources. Terrorist attacks could lead to less oil availability. The development of new nuclear waste storage locations could reduce the cost of using nuclear energy.

The game is structured so the player completes one turn representing choices made during one year, then faces two random events before their second turn begins. These events will impact the safety, cost and environmental impact of particular resources. After these events, the second turn begins. The power needed by the player's city will be significantly higher, and all of the decisions made in the first turn will still be impacting the city. After making decisions to meet the city's new, higher energy needs, the second turn ends. Two more events occur, again changing the costs, security, and environmental impact of each resource. Then, the game is over. The fact that the game is only two turns long means that it takes somewhere between five and fifteen minutes to play it through completely.

The brevity of *Energyville* makes it difficult for it to compete with *Energy City* where players have ten or twenty turns to make changes in the energy infrastructure. *Energy City* is entertaining for multiple playthroughs, as the player gains a broader sense of how to succeed within the game, and because the game has a broader array of possible outcomes. *Energyville* doesn't offer nearly as many ways to succeed or fail, giving it less long-term entertainment value. Because the game itself is so short, this is highly problematic, and the chances of players learning much from the game are lower because the game doesn't require nearly as much learning for players to be successful.

GAMING TAKES PLACE

In his essay on ecocomposition titled "Writing Takes Place," Sidney Dobrin offers the following observation of the ecological nature of writing:

> Writing is an ecological pursuit. In order to be successful, it must situate itself in context; it must grow from location (contextual, historical, and ideological). [. . .] Writers become part of an environment; they are a product of that environment. They write themselves into the order of a system, and they help define that system. [. . .] Writers write from a place, a *topoi*. By writing, they write about that place and define that place. (18–19)

We would substitute "gaming" and "gamer" for "writing" and "writer" here. We contend that game playing is a rhetorical *techne* that creates knowledge through the skills that are required and that emerge through the process of play and through the context laden processes that the player is immersed in both in the game environment itself and the extended context (the classroom, chat rooms, social spheres, and so forth) that the game is played in. To draw from Blair's notion of material rhetoric, although the game's platform, actions, boundaries, and rules are predetermined by the designers, those rules are contingent, and game play itself has real presence and real consequences which may or may not be what the designer rhetors intended. What remains is the degree to which a game design may actively sponsor ecological literacies through its design. We analyzed *Energy City* and *Energyville* to assess each game's potential for such sponsorship.

Energy City begins with a brief tutorial that walks players through their first turn. During this turn, players learn the basic aspects of game play, and the game only adds one element in turn two: the requests of interest groups. Thus, the beginning of the game serves as a useful model of how to play the game. *Energyville*, on the other hand, does much less to prepare the player for success. Only late in the first turn will the player discover petroleum is a required resource, possibly after the player has already used much of their budget. Also, in *Energyville*, players don't have a sense of the overall game structure from the outset. They aren't aware that events will occur after the first turn has been completed, and they don't realize the game is only two turns long. Thus, the game's design is jarring and much less intuitive, like sitting down to watch a movie and having it end twenty minutes later.

Those playing *Energy City* understand the game's length from the outset, and are actually allowed to choose whether the game is ten or twenty turns long. If we think of the game as a narrative, *Energy City* has a clear and consistent structure that helps players to enjoy and immerse themselves in the context of this digital space. *Energyville* is a disjointed narrative that ends surprisingly quickly, before the player even begins to feel comfortable with how the game is played. Due to this jarring structure, *Energyville* fails to create an immersive experience.

Not only will players of *Energyville* find themselves less aware of the game's structure as they play through for the first time, but they'll also interact less with the facts adapted into the game space. Ironically, *Energyville* contains quite a bit of information; however, instead of making it an active part of game play, different items can be clicked on to reveal excruciatingly dense blocks of text. This information remains detached from the actual game play, and can be entirely ignored without harming the gaming experience. Thus, the game is designed to communicate large parts of its information through passive learning instead of asking players to actively engage with this information.

Energy City also allows players to read details about the types of energy included in the game. However, the text is shorter and incorporated more into the game itself. More importantly, aspects of this knowledge must be strategically used in order to succeed. Energy sources that produce less energy than others will actually fill less of the energy bar. If it will cost money to develop a new technology, that cost is listed and must be worked into the player's budget. The level of challenge is high enough that players must think carefully in order to not spend all of their money and lose the game. In *Energyville*, on the other hand, it is very difficult to lose because foolish actions won't actually cause the game to end. Regardless of what the arguments presented by these games are, then, players are more likely to learn the messages communicated by *Energy City*. The game is less confusing than *Energyville*, yet simultaneously more difficult, and also requires players to truly think about their decisions. For these reasons and a variety of others, *Energy City* is a much more entertaining game, and is thus more likely to actually teach players about energy resources.

In terms of meta-level thinking, both games make the effort to create a game space that can inform the real-world cultural models of players. Both of these virtual cities are fairly generic and universal in appearance. The energy sources and reduction techniques are all possibilities that many cities would be capable of utilizing if they chose to do so. Both games, of course, simplify many processes: *Energyville* assumes the city will be in a location where petroleum can be drilled for, nuclear energy can be utilized, and solar, wind, hydroelectric and geothermal energies can all be drawn from the nearby environment. *Energy City* makes unrealistic assumptions as well: it only takes two years to create and begin using most power plants, scientific initiatives always pay off after one year, and outcomes for every kind of resource are predictable.

In this sense, *Energyville* creates a more nuanced understanding of energy resources. The fact that unexpected events can change the availability of resources makes the interaction with particular resources more dynamic in the game. Unfortunately, *Energyville* ignores the complex politics that make it difficult to institute any large-scale changes. *Energy City*, through requiring players to interact with interest groups, allows players to experience a highly simplified version of the conflicts involved in these real-world political decisions.

Through many play-throughs of *Energy City*, the most successful technique seems to be building new energy sources immediately while also initiating conservation projects. This leads to a large up-front cost, but would make it possible to use less fossil fuel a few years later. Conservation methods also slow down the environmental damage that utilizing fossil fuels requires. After a period of five or six years, enough alternative sources of energy could be in place to mostly halt the continued degeneration of air and environmental quality. The conservation initiatives would be actively improving the state of the environment, and the player would have the most difficult stage of the game behind them. Simply continuing to use more sustainable energy sources would be enough to continue improving the environment throughout the duration of the game.

This communicates an argument that holds meaning for real cities as well as virtual ones: immediate changes will have a greater long-term benefit than decisions we put off for even a few years. It also communicates that conservation initiatives can have a real impact on the amount of energy consumed. A plethora of options available can save money in the long term and reduce the need to consume energy. The role of interest groups in the game suggests the fact that different individuals have different perspectives on environmental issues, and that politicians must work to keep them all happy. Beyond suggesting this complexity to the player, the game doesn't make a clear argument about whether the effect of these groups is positive or negative, simply making it clear that these sorts of groups exist and must be taken into consideration.

Energy City's rhetorical argument glosses over the complexity of really changing from one set of energy resources to another, making the argument superficial in a way that isn't openly acknowledged within the game. Within the game, it seems as though any energy source is just as effective as any other. If enough energy can be produced via wind turbines, residents can drive their cars to work without using gasoline. This simplification overlooks the much more complex real-world situation where cars are generally designed for one kind of fuel—or two, at the most.

The arguments made by *Energyville* have a much greater focus on the dynamic nature of these resources in the future. Events can change the price of petroleum dramatically. Wind technologies could become more cost-effective than they are at the beginning of the game. The player must plan for a dynamic, changing set of circumstances. However, during none of the play-throughs of the game did petroleum itself become an optional technology. Despite the changing nature of energy in the game, the continued need to rely on petroleum appears unavoidable.

The between-turn cinematic screens show images of what the city looks like at each stage of its development. Regardless of the player's success or failure at improving the environment, the final incarnation of this city is a pristine, futuristic landscape. Thus, in a game where players are asked to focus on thinking about environmental costs, the game environment is as much an unchanging backdrop as in most other video games. There is really no way to make the environment degenerate. The environmental damage is invisible.

This is a fascinating aspect of both games: neither cityscape ever appears polluted, and the nature around each city remains immaculate. Regardless of how miserable the environment and air quality has become according to the game's meters, the fields and rivers around the cities are unchanged. *Energyville*'s water isn't impacted by the offshore drilling taking place there. The question of environmental impact, then, remains just as abstract and esoteric throughout these games as it is for many city-dwellers who cannot see the long-term changes their actions cause. Why, then, do games that ask us to examine the effects of our energy use not actually confront us with the consequences of environmental neglect?

The lack of a genuinely dynamic environment may reflect a design over-sight, or it may only reflect the limited budgets that designers of educational games are restricted by. It would be much simpler to create a city with a mostly static appearance and reflect the majority of changes in a few barometers around the edges of the screen. Whatever the reason, it is prob-lematic that these games about sustainability don't actually make visual the process of environmental change. Video games have the potential to illustrate the consequences of our treatment of the environment. However, designers seem unaware of this potential and neglect to create a dynamic—and potentially emotional—sense of what environmental degeneration will actually look like.

The structures of both games also create a false sense of isolation and independence for these cities. Financially, politically, and environmentally, the urban place is disconnected from anything beyond its borders, fully capable of financing environmental changes without assistance, and within reach of enough energy sources to support a growing population. Despite the global nature of environmental concerns, both games fail to raise ques-tions about the interconnectedness of urban areas with so-called natural places, or even with nearby cities. Both games, then, serve as introductions to the complexities of energy production without raising questions about interrelated issues like globalization or consumerism.

Both games are partially successful in creating digital representations of real-world problems. They are both able to place these problems in contexts that reflect some level of nuance, although neither game makes its shortcom-ings and simplifications as explicit as it should. In terms of playability, the most significant problem with these games is the fact that neither is able to fully move beyond the very limited contexts of explicitly educational games: both are designed to be quite short and include some kind of narration that draws attention to the educational aspects of the game. Both games can be mastered at the most difficult level after a couple of hours. By contrast, suc-cessful commercial video games take many hours to play to conclusion and include a broad range of difficulty levels, allowing players to revisit them for dozens—or hundreds—of hours' worth of entertainment. These games often do teach players about playing the game and how the game world functions. This teaching is consistently done within a narrative that is strong enough to draw focus away from the actual "work" of learning. The majority of the learning takes place within the game itself, something *Energy City* is much more successful at accomplishing than *Energyville*.

CONCLUSION

The number of video games developed with sustainability-related themes has been growing, and the potential of this medium for both educational and rhetorical purposes has yet to be seen. This genre is developing in an industry where many of the most popular games—*World of Warcraft* and *Everquest II*, for instance—encourage players to conspicuously display their acquisitions (Rettberg 20; Ducheneaut 98). For example, players in *World of Warcraft* spend a great deal of time wandering around in forests, fields, and deserts trying to get a certain number of specific resources through mining, hunting, fighting, or fishing. However, the game is designed to provide a limitless number of these items. Each time a creature is killed or an ore vein is mined, it will reappear after a matter of minutes. The effect this rhetoric of consumption has on players—if the rhetoric has an impact at all—is unknown. Some scholars have even suggested that conspicuous consumption in a virtual locale will reduce the compulsion to do the same in the real world (Bainbridge). Or perhaps this conspicuous consumption will provide new spaces for power dynamics to develop without eliminating the same problematic behavior in the real world. These new spaces/places of conspicuous consumption might simply reinforce the sense of urgency we have to consume.

Nature is depicted in most massively multiplayer online games as an inexhaustible dispenser of resources. This reinforces a demonstrably false yet frequently held belief about our real-world relationship with the environment. A major challenge of sustainability-focused video games, which exist in a larger context where many games glorify consumption, will be to depict the environment as a fragile balance and not a static background. *Energy City* and *Energyville* both attempt to present arguments with some level of nuance, but both neglect to truly present the player with a dynamic environment. The consequences of failing to protect the environment remain only a number on a meter.

Some significant obstacles must be overcome before sustainability-focused video games are likely to reach large audiences. First, they must be designed with learning principles in mind that will make the games as challenging and entertaining as their competitors. Second, they must portray the environment as an interactive spaces and places, not an unchanging backdrop that events happen upon. Third, they must meet these goals while incorporating factual information about real-world issues, and this information should be acquired through active and not passive learning. Until these games are able to meet these goals, they are not likely to be perceived as more than a classroom activity. If video games are going to be educational and convey intentional rhetorical messages about how we live in the world, then gamers must be drawn to them because of genuine interest. Designers must think beyond the confines of classrooms, and must genuinely think about what has been proven to work and not work in the video game medium.

NOTES

1. *PowerUp* (www.powerupthegame.org) take place on the fictional world of Helios, which is on the verge of environmental collapse. "Players take on the role of Engineers, working together designing and building energy solutions to save the world."
2. *Energy City* was designed by the JASON Project, an organization that makes a variety of games focused on science education. *Energyville* was created by The Economist Intelligence Unit and made available online by Chevron. Thus, one of the games was designed by a company that specializes in educational games, and the other was created by economists and promoted by a fossil fuel corporation. This difference in perspective reveals itself in some of the features of game play, which will be discussed later in this article.

WORKS CITED

Aarseth, Espen. *Cybertext: Perspectives on Ergodic Literature.* Baltimore: Johns Hopkins UP, 1997. Print.
Age of Empires. Ensemble Studios. CD-ROM. 1997.
Assassin's Creed II. Ubisoft Montreal. X-Box 360. 2009.
Bainbridge, William Sims. "Virtual Sustainability." *Sustainability* 2 (2010): 3195–3210. Print.
Blair, Carole. "Contemporary US Memorial Sites as Exemplars of Rhetoric's Materiality." *Rhetorical Bodies.* Ed. Jack Selzer and Sharon Crowley. Madison, WI: The U of Wisconsin P, 1999. 16–57. Print.
Bogost, Ian. *Persuasive Games: The Expressive Power of Video Games.* Cambridge, MA: MIT P, 2007. Print.
City Rain. Mother Gaia Studio. Mac OS X Download. 2010.
Dobrin, Sidney I. "Writing Takes Place." *Ecocomposition.* Ed. Christian R. Weisser and Sidney I. Dobrin. Albany: State U of New York P, 2001. 11–25. Print.
Ducheneaut, Nicolas, and Nicholas Yee. "Collective Solitude and Social Networks in *World of Warcraft.*" *Social Networking Communities and E-Dating Services: Concepts and Implications.* Ed. C. T. Romm and K. Setzekorn. Hershey, PA: IGI Global, 2008. Print.
Energy City. The JASON Project. Filament Games. 2009. Web. 2 Jan. 2013.
Energyville. The Economist Group. 2007. Web. 17 Dec. 2012.
Everquest II. Sony Online Entertainment. CD-ROM. 2004.
F1 Race Stars. Codemasters Birmingham. X-Box 360. 2012.
Fallout 3. Bethesda Softworks. CD-ROM. 2008.
Gabriel Knight: Sins of the Fathers. Sierra On-Line. CD-ROM. 1993.
Gabriel Knight II: The Beast Within. Sierra On-Line. CD-ROM. 1995.
Gee, James Paul. *What Video Games Have to Teach Us about Learning and Literacy, Revised and Updated Edition.* New York: Palgrave Macmillan, 2007. Print.

Grand Prix. Activision. Atari 2600. 1982.

Hayes, Elisabeth R., and Sean C. Duncan. *Learning in Video Game Affinity Spaces*. New York: Peter Lang, 2012. Print.

The Jason Project. "About JASON." *The Jason Project*. 2012. Web. 31 Dec. 2012.

Libby, Brian. "Sustainability-Themed Computer Games Come to the Classroom." *Edutopia*, 28 Jan. 2009. Web. 8 Dec. 2012.

McGonigal, Jane. *Reality is Broken: Why Games Make Us Better and How They Can Change the World*. New York: Penguin, 2011.

Mike Tyson's Punch-Out! Nintendo Research and Development Team 3. Nintendo Entertainment System. 1987.

MiniMonos. MiniMonos Ltd. 2009. Web. 1 Dec. 2011.

Murray, Janet. *Hamlet on the Holodeck: The Future of Narrative in Cyberspace*. New York: Free P, 1997. Print.

NBA 2K13. Visual Concepts. X-Box 360. 2012.

Plan It Green. Masque Publishing. CD-ROM. 2009.

PowerUp. IBM, TryScience/New York Hall of Science. Mac OS X Download. 2008.

Rettberg, Scott. "Corporate Ideology in *World of Warcraft*." *Digital Culture, Play and Identity: a World of Warcraft Reader*. Ed. Hilde Corneliussen, Jill Walker Rettberg. Cambridge, MA: MIT P, 2008. Print.

The Secret of Monkey Island. LucasArts Games. CD-ROM. 1992.

Sherlock Holmes: Nemesis. Focus Home Interactive. CD-ROM. 2008.

Sid Meier's Civilization. MicroProse. CD-ROM. 1991.

SimCity. Nintendo Entertainment Analysis and Development. Super Nintendo Entertainment System. 1991.

Space Jam. Sculptured Software. Playstation. 1996.

Star Wars: Knights of the Old Republic. LucasArts Games. CD-ROM. 2005.

Super Mario Kart. Nintendo Entertainment Analysis and Development. Super Nintendo Entertainment System. 1992.

World of Warcraft. Blizzard Entertainment. CD-ROM. 2004.

World Without Oil. Electric Shadows. Web. 2007.

Part III

Places We Travel Through, Around, and Within

9 Reading the *Atlas of the Patagonian Sea*
Toward a Visual-Material Rhetorics of Environmental Advocacy

Amy D. Propen

In 2009, the conservation organizations Birdlife International and Wildlife Conservation Society (WCS) partnered to produce a digital text called the *Atlas of the Patagonian Sea: Species and Spaces.* The *Atlas* was created to aid in policy decisions related to fisheries management and the designation of transportation routes within the Patagonian Sea, which ranges from southern Brazil to southern Chile, and is increasingly threatened by development and overfishing. Significantly, the *Atlas* is also the first-ever such work created largely by tracking the movements of seabirds, marine mammals, and sea turtles in the area. These marine species were outfitted with remote tracking devices for a ten-year period, and the resulting satellite data was then used to compile the *Atlas.* In this chapter, I analyze the *Atlas* and, in doing so, advocate for an approach to environmental rhetoric that is located at the intersections of animal studies,[1] critical cartography,[2] and what I call visual-material rhetorics, an approach to visual rhetoric attuned not only to the persuasive components of visual artifacts but also to the ways in which considerations of space, bodies, and materiality specifically influence the rhetorical analysis of the visual (Propen). I explore the intersections of these fields by considering how cartographic practice, when understood as a rhetorical knowledge-making endeavor, can help advocate for nonhuman animal bodies residing in the spaces represented through the map. Drawing on Michel Foucault's theory of heterotopias and Carole Blair's theory of material rhetoric, I first situate rhetorical objects as not merely textual but also as visual and material—as spatial. I then analyze the *Atlas* to show how, as an object of visual-material rhetoric, it advocates for species conservation and, in the process, illuminates questions about the use and value of the nonhuman animals who both contributed to the *Atlas*'s creation and are represented through it.

For Carole Blair, material rhetoric involves looking beyond a text's symbolic meaning to also consider its broader consequences on bodies in the world. In redefining what gets to count as a text, Blair considers the impacts and consequences that texts can have in the rhetorical situation. To this end, she questions "the significance of the text's material existence," its "degrees of durability [and] modes or possibilities of reproduction or preservation," as well as how the text works "with, or against" other texts (30). Perhaps most important, Blair is interested in how the text acts on bodies in the world (30). Whereas Blair refers to the text's impacts on people specifically, we might extend her inquiry to consider the text's influence on nonhuman bodies as well. Based on these ideas, I understand "text" not only in the more traditional sense of printed words on the page, but also as multimodal, or as combining visual, material, auditory, or digital information. Many forms of visual rhetorical texts, like memorials and maps, are also material, in that they invite a fuller engagement with the text that transcends a passive reading. Cartographic practice, when considered both a rhetorical endeavor and a conservation effort, can provide a clear lens through which to understand the role that visual-material artifacts can play in environmental and wildlife conservation efforts. In particular, conservation mapping is one area of cartographic practice that implicitly understands mapping as a rhetorical activity compatible with the considerations of visual-material rhetoric.

Conservation mapping refers to spatial practices that designate "geographical areas as relevant for conservation" (Harris and Hazen 52). The reliance on maps to designate conservation areas may inadvertently perpetuate the idea that human and nonhuman animals should be viewed as separate, overemphasize boundaries, or perpetuate "an overly-fixed and static approach to conservation" (56–57). Conservation mapping, like critical cartography, understands mapping as always already "reflective of, and productive of, power" (63).

To view the map as power-laden, fraught with multiple meanings in tension, or representative of heterogeneous, contested spaces is compatible with Michel Foucault's notion of heterotopias, or heterotopic space. Foucault states that we all reside in heterogeneous spaces—that "we do not live in a kind of void, inside of which we could place individuals and things"; rather, "we live inside a set of relations that delineates sites which are irreducible to one another and absolutely not superimposable on one another" (23). We can try to characterize these sites by "looking for the set of relations by which a given site can be defined" (23). Such sites have common but multiple uses, and are frequented by various bodies with various sets of sometimes common, sometimes competing goals. In particular, Foucault is interested in those sites "that have the curious property of being in relation with all the other sites, but in such a way as to suspect, neutralize, or invert the set of relations that they happen to designate, mirror, or reflect" (24). I argue here that the map counts as a version of heterotopic space, as it is a representation of a particular "real" territory, and often conveys multiple arguments or ideas about a place. Such ideas are often borne out of knowledge claims that result in competing or contested discourses about what counts as the most "accurate" representation of a single territory. Maps can then be understood as both visual-material and heterotopic rhetorical artifacts. To pair Foucault's theory of heterotopias with Blair's theory of material rhetoric allows us to understand rhetorical objects as not merely textual but also visual and material—as spatial. Moreover, to understand heterotopias as materially rhetorical allows for an understanding of heterotopic space as influencing the bodies inhabiting that space. With these ideas in mind, this essay takes the *Atlas of the Patagonian Sea* as its primary object of analysis, with a specific focus on the question of how the act of mapping implicitly portrays particular species and places, and how such mapping practices, in producing and perpetuating rhetorical, heterotopic representations, can impact the lives and futures of the nonhuman animals represented through it.

READING THE ATLAS OF THE PATAGONIAN SEA

The *Atlas* is a complex heterotopic representation of this marine ecosystem and the species that reside within it. Available to the public online at http:// atlas-marpatagonico.org, the site's message is presented through a combination of text and images that work together to argue for the conservation of the Patagonian Sea and its marine species. The work contains two main sections: "Spaces" and "Species," along with nine additional links that appear at the bottom of every page within the site. Five of these nine pages provide context related to the Spaces and Species sections, and are titled "The *Atlas*," "The Maps," "GIS," "Threats to Biodiversity," and "Important Marine Areas." The other four pages provide copyright and supporting information. Because it is beyond the scope of this essay to analyze the *Atlas* in full, I focus specifically on the first page of the Species section, "Wandering Albatross," and material from the GIS page that speaks to the relationships between marine mammals and remote satellite technologies. Finally, I invoke ideas and concepts from throughout the rest of the *Atlas* as it is relevant to my overall discussion or helps to contextualize points made in the *Atlas* itself.

The Species Section

The Species section describes the "use of the Patagonian Marine Ecosystem by seabirds and marine mammals" (Falabella, Campagna, and Croxall "Home"). The editors implicitly convey an ecological perspective that understands the natural world as a system in which humans, wildlife, and habitats are interconnected when they note that "[t]hese top predators require considerable space and resources and are vulnerable to many human activities. This makes them good indicators of the status of conservation of the ecosystem" ("The *Atlas*"). In this view, the ecosystem is not seen as separate from the needs of marine mammals and seabirds, nor is it seen as impervious to the impacts of human use.

The marine species in this section fall within the five main categories: albatrosses, petrels, penguins, pinnipeds (seals and sea lions), and marine turtles. Each category contains between one and five species. For example, listed under the first category of "Albatross" are the Wandering Albatross, Northern Royal Albatross, Black-browed Albatross, Grey-headed Albatross, and Light-mantled Albatross. Each species is given its own page, which typically includes two to five visual representations (such as photographs and maps) and one textual description of the species being described. I describe the Wandering Albatross page here, to provide a sense of how pages in this section are organized and the arguments they make.

The Wandering Albatross page contains a photograph[3] of the albatross, followed by four maps that describe nesting sites, principal feeding areas, and distribution and usage data from spring to summer, and autumn to winter, respectively (Falabella et al. "Wandering Albatross").

The photo depicts a portrait of the Wandering Albatross in flight above the ocean (Suter). Clicking on the image enlarges the photo, revealing the whole of the bird's body as it soars above the ocean, thus making salient its impressive wingspan. The albatross is pictured in its natural environment, and through the photo's composition, which provides a view of the bird soaring over the ocean with its wings fully extended, viewers get a sense of the bird's uniquely impressive aesthetic characteristics. For Quinn R. Gorman, nature photographs are rhetorical artifacts that help shape "the cultural meaning and categories" that subsequently allow humans "to take substantive, positive steps toward change. We can only act based upon our understanding of what nature is and how we might protect it," even if those understandings are based on culturally constructed representations of reality (254). The photo of the Wandering Albatross affords its viewer a mental picture of what might otherwise be an abstract concept of this bird, thus providing a clearer visual understanding of the species that needs protecting, even if the photo in itself constitutes a cultural construction. To the right of the photograph viewers will then find the nesting sites map (see Figure 9.1).

The purpose of this map is to represent the populations of Wandering Albatross that nest and breed in the Patagonian Marine Ecosystem. Latitudes and longitudes are visible along the map's x and y axes, respectively, and a legend in the top left corner orients readers toward North and notes a scale of one centimeter to 340 kilometers. Landforms and surrounding ocean are shown in grayscale, presumably in order to emphasize the use of green and blue to communicate information about albatross populations studied and represented. Consistent with this reading and the information conveyed in the map's caption, the bright green dot and blue text represent the nesting colonies of Wandering Albatrosses on the Island of South Georgia[4] and the "percentage of the world population that each area represents relative to the total number" (Falabella et al. "The Maps"). Blue text also refers to "populations for which the *Atlas* provides distribution data"; thus, the blue text that reads "Is. Georgias del Sur: 18%," in combination with the green dot, mean that the *Atlas* provides distribution data related to populations of Wandering Albatross residing on the Island of South Georgia.

Sitios de nidificación regional del albatros errante (*Diomedea exulans*)
Regional Nesting Sites of the Wandering Albatross (*Diomedea exulans*)

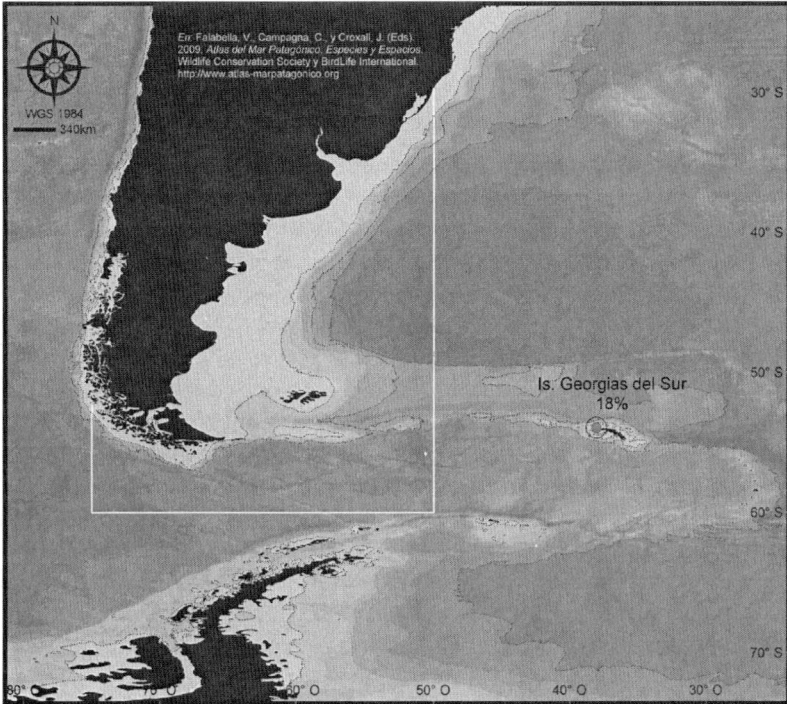

Figure 9.1   Regional nesting sites of the Wandering Albatross (Diomedea exulans). (Source: Falabella et al. "The Wandering Albatross.")

Further, the text to the right of the map notes that the "world breeding population" of Wandering Albatrosses is "estimated at only 8,050 pairs" and that the species was designated "vulnerable" in 2008 by the International Union for Conservation of Nature; thus, if the Wandering Albatross colonies on the Island of South Georgia represent eighteen percent of the world population of Wandering Albatross, and the world breeding population is estimated at 8,050 pairs, the *Atlas* provides distribution data for approximately 1,149 pairs of Wandering Albatross, or the eighteen percent of the world's population that nest on the Island of South Georgia. The blue circle around the green dot represents the "colonies of origin" of the Wandering Albatrosses studied "with remote tracking devices." In other words, the Wandering Albatrosses studied and represented in the *Atlas* originated in colonies in and around the Island of South Georgia (Falabella et al. "The Maps," "Wandering Albatross").

Following the nesting sites map are three maps that show the principal feeding areas and seasonal usage areas (spring–summer and autumn–winter) for the Wandering Albatross. The purpose of these maps is, ultimately, to demonstrate that the Continental Shelf and Continental Slope (areas of ocean closer to and farther from the coast, respectively) are important habitats for marine species throughout the year, regardless of seasonal variation or migration and nesting patterns. To demonstrate the albatrosses' reliance on the shelf and slope regions, researchers used the GIS data gleaned from tracking the migratory and foraging trips made by these birds to create what are called "density distributions and utilisation distribution contours" that "indicate the distribution range of the animals studied and the areas with similar probabilities of occurrence" (Falabella et al. "The Maps"). The researchers define Geographical Information Systems, or GIS, as "a set of tools that allows geographically referenced information to be stored, edited, analysed, integrated, shared and displayed" (Falabella et al. "GIS"). Using GIS, researchers were able to take remote tracking data from the birds' migratory and foraging trips, and from that data make estimations about patterns of usage and where the birds are most and least likely to congregate. These estimations can be converted using what is called the "kernel density" feature in GIS, to create maps that show broader density distribution and usage patterns ("Kernel Density"). In the case of the maps in the *Atlas*, utilization distribution contours are represented at intervals of fifteen, seventy-five, and ninety-five percent density. As the researchers describe,

> The darkest areas (50%) identify the zones where individuals remained for the longest time. Therefore, the 50% contour reveals the areas of highest density or probability of occurrence (they do not necessarily indicate the zones where individuals feed, but do indicate areas of intense use). As we expand the area of the contour to the 75% or 95% zones, the distribution range where the probability of occurrence is lowest can be seen. Lines indicate the complete distribution range (100%). (Falabella et al. "The Maps")

In the principal feeding areas map (see Figure 9.2), it is clear that the Continental Slope and the Argentine Basin comprise the primary feeding area, or the area of most "intense use," for the Wandering Albatross. These regions are highlighted in dark teal, or the fifty percent contour, as the legend indicates, and thus represent the area of highest density for the albatross. The text in the map's caption supports this reading, stating: "The slope and the Argentine Basin are the principal feeding areas for the Wandering Albatross from South Georgia year round. Unlike other albatrosses, there is little use of shallow waters on the continental shelf" (Falabella et al. "Principal Feeding Areas"). The caption also invokes the use of remote tracking technology to monitor marine species, noting that the data used to create this map is based on 105 trips made by both male and female Wandering Albatrosses residing on the Island of South Georgia.

Albatros errante (*Diomedea exulans*): Áreas de alimentación anual en el Mar Patagónico
Wandering albatross (*Diomedea exulans*): Principal feeding areas in the Patagonian Sea

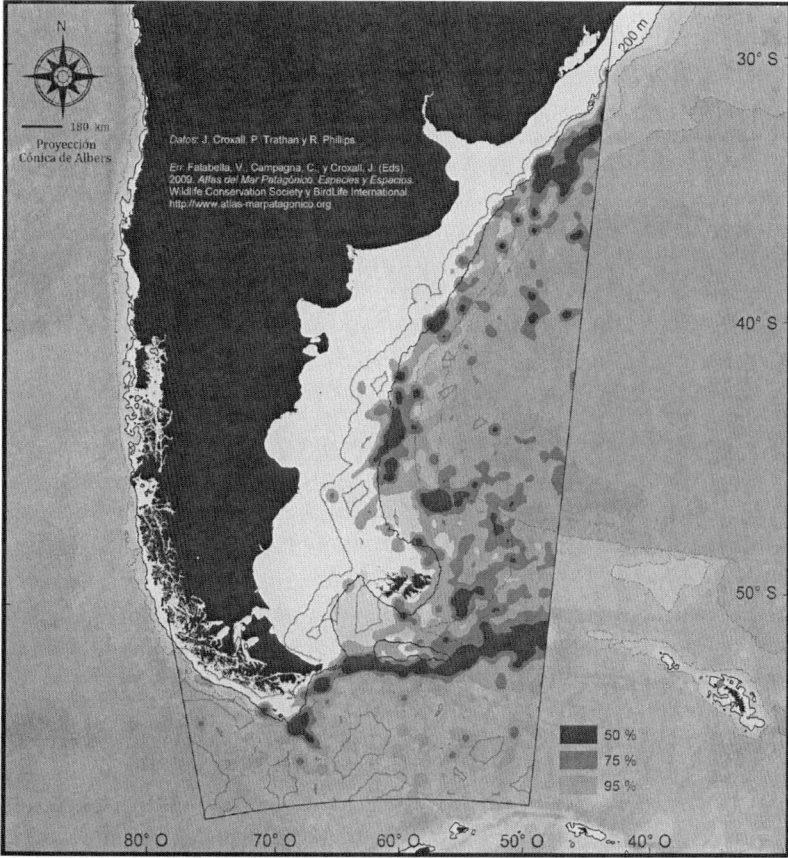

Figure 9.2 Wandering Albatross (Diomedea exulans): principal feeding areas in the Patagonian Sea. (Source: Falabella et al. "Principal Feeding Areas"; Dataholders: J. Croxall, P. Trathan and R. Phillips.)

The maps on the Wandering Albatross page provide evidence for this seabird's heavy reliance on the Continental Slope. The seasonal usage and distribution patterns in the spring–summer and autumn–winter maps continue to depict heavy usage of the Continental Slope by both sexes of Wandering Albatross. Combined with photographs of the albatross, data about its vulnerable status, and additional text that describes its primary diet of squid and fish, noting also, for example, that albatrosses will "follow vessels frequently, competing with other species to take advantage of fishing discards," the elements of the page together convey not only the aesthetic value of the Wandering Albatross and the importance of the ecosystem for this species, but also the precarious status of the species as well as its risky interactions with and reliance on the fishing industry (Falabella et al. "Wandering Albatross"). In other words, the Wandering Albatross page functions as a representation of the heterotopic, contested space that this species must negotiate.

Researchers converted data about the albatrosses' trips into maps that show the areas of ocean most valuable to them. From the perspective of material rhetoric, the birds' movements can be recorded and reproduced in the form of visual-material rhetorical artifacts that can influence conservation policy in the region. Also implicit in the principal feeding areas map is the idea that the maps themselves would not be possible without the use of remote tracking technologies to monitor and record the trips made by these marine species.

REMOTE TRACKING TECHNOLOGY, VISUAL-MATERIAL ENVIRONMENTAL RHETORICS, AND THE CONSERVATION OF MARINE SPECIES

Perhaps the most remarkable aspect of the *Atlas* pertains to the manner in which much of its cartographic data was collected. Again, the *Atlas* is the first of its kind to use remote tracking technologies to monitor and record the trips made by species of marine mammals, marine turtles, and seabirds in the region. Whereas the maps in the Spaces and Species sections clearly provide the foundation upon which the document as a whole is built, the page describing the researchers' use of GIS technology contains some of the *Atlas's* most rhetorically powerful imagery related specifically to the remote tracking of marine species and their involvement in this mapping endeavor.

The GIS database used to create the density maps houses "285,000 localisations, describing thousands of migratory and foraging trips of seabirds, marine mammals and sea turtles" (Falabella et al. "GIS"). Researchers used a variety of remote tracking technologies to record the trips of these marine species: "Most of these data were obtained by satellite transmitters (PTT), although geolocators (GLS) and recorders for global positioning systems (GPS) were also deployed." Juxtaposed with this discussion of remote tracking technology is an image of an elephant seal with a satellite tracking device placed on its head (see Figure 9.3).

The image also includes a caption that contextualizes the placement of the device: "At sea, when elephant seals surface to breathe, only the head emerges above water. Thus, the instruments must be placed on it to allow a transmission to the satellites. In other species (such as birds), the back is the most suitable location to deploy transmitters" (Falabella et al. "GIS"). The researchers' brief discussion here of the specific placement of remote sensing devices on the bodies of marine species speaks to the history of remote

Figure 9.3   Elephant seal with satellite tracking device. (Photograph Courtesy of Victoria Zavattieri; Falabella et al. "GIS.")

sensing technologies in wildlife research, and reflects shifting approaches in its use.

Discussions of tracking wildlife for conservation gained ground in the early 1970s, around which time legislation like National Environmental Policy Act of 1969, the Marine Mammal Protection Act (MMPA) of 1972, and the Endangered Species Act of 1973 began emphasizing "science and technology as the most promising means of mitigating the effect on wild animals of growing human populations and levels of consumption" (Benson 2). Initial attempts in the 1960s to radio-tag dolphins explored "the relationship between vocalizations and behavior" in marine mammals, although the technology at that point made such efforts "barely feasible" (140). Early tracking technologies like acoustic tags were eventually ruled out "because of concerns about the animals' dependence on sound for communication and navigation" (142). Additionally, marine mammals often required capture in order to attach the devices. Later developments in the early 1990s made tagging more feasible and "non-intrusive," as newer tags could be "attached to free-swimming whales by suction cup" (174–175).[5] Recent advances in satellite telemetry have again shifted understandings of the technology and its uses: "Instead of being seen as experts whose technologically mediated intimacy with wild animals gave them authority to speak on their behalf, scientists could now be seen as mediators of a kind of virtual intimacy between individual animals and mass audiences, or even as audiences themselves" (190). As readers of the *Atlas* view the photograph of the elephant seal with the satellite tracking device, for example, they may very well feel a curiosity about this creature's life, its relationship to humans, and the knowledge made possible through its technologically mediated body.

The photo of the seal undoubtedly invites reflection on the part of the viewer. The seal is clearly part of the natural world—seemingly at home in its physical environment and unaware of the device it carries. At the same time, the device constitutes an intervention into its existence—its body mediated by remote tracking technology. On the one hand, the photo reflects what Cary Wolfe describes as a "humanist posthumanism," as it "reinstalls a very familiar figure of the human at the very center of the universe of experience . . . or representation" (148). The photo foregrounds the human technological intervention in the life of this marine mammal, thus placing the human, or a representation of human action—the ability to track—at the center of the image. On the other hand, the photo also posits what Wolfe refers to as a more posthumanist version of posthumanism—one that "take[s] seriously the ethical and even political challenges of the existence of nonhuman animals" (148). That is, by situating the technology as able to help advocate for marine species, the photo implicitly communicates the challenges to these species' continued existence and the measures being taken to allay those threats. In either case, however, the technological mediation of the seal's body intervenes in the seal's autonomy as an independent creature. This idea raises questions about how humans conceptualize the autonomy of nonhuman animals—a consideration central to those who study human/nonhuman animal relations.

Marcus Bullock considers that humans need nonhuman animals more than nonhuman animals need us; as he puts it,

> while we may be only an inessential part of their habitat, they are an indispensable element of ours . . . [A]s long as we remain out of sight and sound and scent, they do not think about us . . . We are simply a disturbance in their lives that would have no existence if it ceased to occur. (106–107)

Bullock states that scientific observation has revealed many similarities between human and nonhuman animals, "[a]nd yet the very study that permits us to learn of this similarity also establishes it as a profound form of difference between them and us" (107–108). Thus, he says, we continue to understand "the world of an animal" as different from "the world that we construct for ourselves" (107–108). Moreover, he suggests, we require this sort of understanding in order to justify our relationship with the excesses that make us human: "Our sensory apparatus burdens us, as reflexively conscious beings fully aware of time and change, with the inescapable demand that we bring that excess to order by all the conflicting means of language" (107–108). In this view, the satellite tracking device imposes a human technological agenda on to nonhuman animals—one that privileges the function of epistemic sensory apparatuses to better understand the world of a perceived Other. Bullock also says, however, that humans would be remiss in perpetuating "the rigidly assured vision that sees nothing but the operation of human knowledge anywhere in the universe. The steadfast refusal to see expressiveness anywhere merely becomes another species of anthropomorphism," one that insists on "hearing only silence and seeing only empty matter in the language of animal forms" (112). On the one hand, the technological mediation of the seal's body, as represented through the photo, constitutes a human intervention into the animal's life and bears implications for the extent to which we perceive nonhuman animals as autonomous beings; on the other, such mediations also allow humans to engage in new forms of expressiveness—ones that move audiences to perceive nonhuman animals from ecological perspectives that allow for the co-construction of new knowledge-making and conservation practices.

Conservation mapping projects such as this one speak to a cultural moment in which tracking technologies are perceived as inviting more specific and interactive ways of understanding and engaging with nonhuman animals. The photo of the seal, and the act of tracking represented therein, advocate for a curiosity about the physical world around us—specifically, a curiosity about the bodies residing in those worlds, and a curiosity rooted in the desire to make that world a better place. As Nigel Rothfels says, "the stakes in representing animals can be very high . . . [How] we talk or write about animals, photograph animals, think about animals, imagine animals—represent animals—is in some very important way deeply connected to our cultural environment" (xi). In highlighting a cultural moment in which tracking technologies are perceived as able to foster stronger engagement with and influence on marine species, the photo functions not only as an artifact of visual-material, environmental rhetoric but also reflects a particular rhetoric of technology. As Charles Bazerman describes, "technology, as a human-made object, has always been part of human needs, desires, values, and evaluation, articulated in language and at the very heart of rhetoric" (383). He thus understands the rhetoric of technology as

> the rhetoric that accompanies technology and makes it possible—the rhetoric that makes technology fit into the world and makes the world fit with technology. There is a dialectic between rhetoric and the material design as the technology is made to fit the imaginably useful and valuable, to fit into people's understanding of the world. (385)

Given the views of Rothfels and Bazerman, the photo of the seal and the use of remote tracking technology in conservation mapping can be understood as reflecting current cultural discourses of what counts as legitimate communication about environmental issues. Moreover, the photo of the seal reflects a human desire to better understand the worlds in which we live and our relationships to those species with whom we share it. The photo also communicates a narrative about the satellite tracking device—one in which humans and marine species can interact by way of a technological mediation that potentially allows for new understandings of the ecosystems shared by the two. The seal becomes a co-author of a map of its own existence; the researcher becomes, as Benson has suggested, a mediator of a sort of "virtual intimacy" between human and nonhuman animals. Such a narrative makes it possible to imagine how this technology fits into the world—and how the world fits with the technology at this particular moment in time. Such forms of technological discourse thus represent the interplay between discourse and the specificity of the cultural moment—they represent a specific "kairos" (Miller 83). The photo of the seal with the tracking device constructs a narrative related to how the technology fits with this specific rhetorical situation at this specific time and place. In doing so, the image represents the broader rhetorical work of the *Atlas* and the particular cultural moment in which it is immersed.

CONCLUSION

As an artifact of visual-material, environmental rhetoric, the *Atlas* is also a rhetoric of technology that reflects the shifting contexts that influence approaches to wildlife conservation. It is a complex heterotopia whose implicit and explicit arguments reflect the competition for resources within an ecosystem shared by human and nonhuman animals. As a result, the *Atlas* draws on rationales for conservation that are rooted in combinations of ethical, utilitarian, ecological, and aesthetic attitudes. An analysis of the *Atlas* also demonstrates how visual-material rhetorics can be attentive to issues involving cartographic practice, environmental advocacy, and non-human bodies.

The use of remote tracking technologies to help create the *Atlas* reflects not only contemporary understandings of what counts as culturally credible conservation mapping practices, but also a perceived need to better understand the lives of marine species and our coexistence with them in a threatened ecosystem. Harris and Hazen ask "what is at stake in defining and mapping protected areas for conservation?" (50). Likewise, Blair questions the significance of the text's existence beyond its immediate role in the rhetorical situation. I argue that how we represent nonhuman animals reflects our cultural understandings of the relationships between humans, nonhumans, the environment, and how technology functions to mediate those relationships. Within the context of the *Atlas*, remote tracking technologies allow marine species to illustrate their own paths and tell their own conservation stories. These stories, and the technological mediation that enables them, also have rhetorical implications for how we understand our relationships to nonhuman animals and how we perceive their status as autonomous beings in the world. Although future work should continue to explore these implications, it is clear that the *Atlas*, as a visual-material rhetorical artifact, illuminates a cultural moment that reveals a perceived responsibility on the part of humans to engage with nonhuman animals in ways that use cartographic technologies to help ensure the generativity of vulnerable species.

ACKNOWLEDGEMENTS

I would like to thank the editors of the *Atlas*: Valeria Falabella and Claudio Campagna at the Wildlife Conservation Society, and John Croxall at Birdlife International, for their permission to reprint the maps included in this essay. Thanks also to Victoria Zavattieri for permission to reprint the photograph in Figure 9.3. Additionally, I would like to thank Valeria Falabella for her helpful clarifications and feedback throughout the writing of this essay, as well as Peter Goggin for his useful comments and suggestions during the writing and editing process.

NOTES

1. The field of animal studies may be defined as "the interdisciplinary study of the relationship between humans and other animals" ("Animal Studies"). Broadly construed, such explorations may focus on questions of animal rights, the function of anthropocentrism in humans' conceptualizations of the use or value of nonhuman species in society, or the ways in which anthropomorphism shapes our understandings of human/nonhuman animal relationships, among, of course, many other possibilities. For a highly comprehensive reading list, see Michigan State University's "Animal Studies Bibliography": http://animalstudies.msu.edu/bibliography.php (Kalof, Bryant, and Fitzgerald).
2. Critical cartography, often considered a contemporary subdiscipline of cartographic studies, understands geographic knowledge as reflective of cultural contexts and power relations (Crampton and Krygier 11). Many scholars who study critical cartography also explicitly understand mapping as a rhetorical practice that emphasizes the "discursive power of the medium, stressing deconstruction, and the social and cultural work that cartography achieves" (Kitchin, Perkins, and Dodge 3–4).
3. Readers may view this photo by visiting http://atlas-marpatagonico.org/species/8/wandering-albatross.htm.
4. It is important to provide a brief clarification about place names. The *Atlas* uses a South American, Argentinean perspective when referring to locations in the Patagonian Sea region. The Island of South Georgia, for example, from a North American perspective, is more commonly known as the British Overseas Territory of South Georgia and the Sandwich Islands. Moreover, the *Atlas* is available in both English and Spanish (the site's home page allows viewers to choose between English and Spanish versions), and many of the downloadable maps in the *Atlas* include captions

with both English and Spanish translations, and often refer to place names in Spanish (e.g. Is. Georgias del Sur). In this essay, I use the South American, English-language version of the place name (Island of South Georgia), except when quoting directly from a map that notes the place name in Spanish (Is. Georgias del Sur).

5. The researchers' protocol for attaching the satellite tracking device to the seals tracked for the *Atlas* project did require some mode of human intervention; however, the seals were not physically harmed in any way, and efforts were made to minimize their disturbance. In brief, "[t]he procedure involved approaching a sleeping or resting seal, injecting the drug and retreating quickly, requiring less than 1 min to perform. The behavior of the subject was then observed at a distance to decrease disturbance during induction. Immobilizations were done at low tide as a safety measure to prevent seals from moving to the ocean during the induction phase" (Campagna et al. 1795). The tracking devices were then "attached on the head with marine epoxy . . . to facilitate transmission of location as the animal surfaced between dives. Immobilizations were smooth and uneventful; all animals behaved normally within 2 h of the injection" (1795).

WORKS CITED

"Albatross *Diomedeidae.*" *Animals: National Geographic Wild*. National Geographic Society. 2011. Web. Nov. 2011.

"Animal Studies at MSU." *Animal Studies at Michigan State University*. Michigan State University. 2011. Web. Nov. 2011.

Bazerman, Charles. "The Production of Technology and the Production of Human Meaning." *Journal of Business and Technical Communication* 12.3 (1998): 381–387. Print.

Benson, Etienne. *Wired Wilderness: Technologies of Tracking and the Making of Modern Wildlife*. Baltimore: Johns Hopkins UP, 2010. Print.

Blair, Carole. "Contemporary U.S. Memorial Sites as Exemplars of Rhetoric's Materiality." *Rhetorical Bodies*. Ed. Jack Selzer and Sharon Crowley. Madison: U of Wisconsin P, 1999. 16–57. Print.

Bullock, Marcus. "Watching Eyes, Seeing Dreams, Knowing Lives." *Representing Animals*. Ed. Nigel Rothfels. Bloomington: Indiana UP, 2002: 99–118. Print.

Campagna, Claudio, Alberto R. Piola, Maria Rosa Marin, Mirtha Lewis, Uriel Zajaczkovski, and Teresita Fernández. "Deep Divers in Shallow Seas: Southern Elephant Seals on the Patagonian Shelf." *Deep Sea Research Part I: Oceanographic Research Papers* 54 (2007): 1792–1814. Print.

Crampton, Jeremy W., and John Krygier. "An Introduction to Critical Cartography." *ACME: An International E-Journal for Critical Geographies* 4.1 (2006): 11–33. Web. June 2009.

Croxall, John, P. Trathan, and R. Phillips, Dataholders. "Principal Feeding Areas." *Atlas of the Patagonian Sea. Species and Spaces*. Ed. Valeria Falabella, Claudio Campagna, and John Croxall. Buenos Aires, Wildlife Conservation Society and BirdLife International, 2009. Web. Nov. 2011. http://atlas-marpatagonico.org/species/8/wandering-albatross.htm

Falabella, Valeria, Claudio Campagna, and John Croxall. "The *Atlas.*" *Atlas of the Patagonian Sea. Species and Spaces*. Ed. Valeria Falabella, Claudio Campagna, and John Croxall. Buenos Aires, Wildlife Conservation Society and BirdLife International, 2009. Web. Nov. 2011. http://atlas-marpatagonico.org/the-atlas.html

————. "GIS." *Atlas of the Patagonian Sea. Species and Spaces.* Ed. Valeria Falabella, Claudio Campagna, and John Croxall. Buenos Aires, Wildlife Conservation Society and BirdLife International, 2009. Web. Nov. 2011. http://atlas-marpatagonico.org/gis-database.html

————. "Home." *Atlas of the Patagonian Sea. Species and Spaces.* Ed. Valeria Falabella, Claudio Campagna, and John Croxall. Buenos Aires, Wildlife Conservation Society and BirdLife International, 2009. Web. Nov. 2011. http://atlas-marpatagonico.org/home-spaces-species.html

————. "The Maps: Interpreting the Maps in the *Atlas*." *Atlas of the Patagonian Sea. Species and Spaces.* Ed. Valeria Falabella, Claudio Campagna, and John Croxall. Buenos Aires, Wildlife Conservation Society and BirdLife International, 2009. Web. Nov. 2011. http://atlas-marpatagonico.org/the-maps.html

————. "Regional Nesting Sites of the Wandering Albatross (Diomedea exulans)." *Atlas of the Patagonian Sea. Species and Spaces.* Ed. Valeria Falabella, Claudio Campagna, and John Croxall. Buenos Aires, Wildlife Conservation Society and BirdLife International, 2009. Web. Nov. 2011. http://atlas-marpatagonico.org/species/8/wandering-albatross.htm

————. "Wandering Albatross: *Diomedea exulans*." *Atlas of the Patagonian Sea. Species and Spaces.* Ed. Valeria Falabella, Claudio Campagna, and John Croxall. Buenos Aires, Wildlife Conservation Society and BirdLife International, 2009. Web. Nov. 2011. http://atlas-marpatagonico.org/species/8/wandering-albatross.htm

Foucault, Michel. "Of Other Spaces." Trans. Jay Miskowiec. *Diacritics* 16 (1986): 22–27. Print.

Gorman, Quinn R. "Evading Capture: The Productive Resistance of Photography in Environmental Representation." *Ecosee: Image, Rhetoric, Nature.* Ed. Sidney I. Dobrin and Sean Morey. Albany: State U of New York P, 2009: 239–256. Print.

Harris, Leila, and Helen Hazen. "Rethinking Maps From a More-Than-Human Perspective: Nature-Society, Mapping and Conservation Territories." *Rethinking Maps: New Frontiers in Cartographic Theory.* Ed. Martin Dodge, Rob Kitchin, and Chris Perkins. New York: Routledge Studies in Human Geography, 2009. 50–67. Print.

"How Kernel Density Works." *ArcGIS Desktop 9.3 Help.* Environmental Systems Research Institute, Inc. (ESRI), n.d. Web. Nov. 2011.

Kalof, Linda, Seven Bryant, and Amy Fitzgerald. "Animal Studies Bibliography." *Animal Studies at Michigan State University.* Michigan State University. 2011. Web. Nov. 2011.

Kitchin, Rob, Chris Perkins, and Martin Dodge. "Thinking about Maps." *Rethinking Maps: New Frontiers in Cartographic Theory.* Ed. Martin Dodge, Rob Kitchin, and Chris Perkins. New York: Routledge Studies in Human Geography, 2009. 1–25. Print.

Miller, Carolyn R. "Opportunity, Opportunism, and Progress: Kairos in the Rhetoric of Technology." *Argumentation* 8 (1994): 81–96. Print.

Propen, Amy D. *Locating Visual-Material Rhetorics: The Map, the Mill, and the GPS.* Anderson, SC: Parlor P, 2012. Print.

Rothfels, Nigel. "Introduction." *Representing Animals.* Ed. Nigel Rothfels. Bloomington: Indiana UP, 2002. vii–xv. Print.

Suter, Claudio. "Wandering Albatross (*Diomedea exulans*)." *Atlas of the Patagonian Sea. Species and Spaces.* Ed. Valeria Falabella, Claudio Campagna, and John Croxall. Buenos Aires, Wildlife Conservation Society and BirdLife International, 2009. Web. Nov. 2011. http://atlas marpatagonico.org/species/8/wandering-albatross.htm

Wolfe, Cary. "From *Dead Meat* to Glow-in-the-Dark Bunnies: Seeing 'the Animal Question' in Contemporary Art." *Ecosee: Image, Rhetoric, Nature.* Ed. Sidney I. Dobrin and Sean Morey. Albany: State U of New York P, 2009: 129–152. Print.

Zavattieri, Victoria. "Elephant Seal with Satellite Tracking Device." *Atlas of the Patagonian Sea. Species and Spaces.* Ed. Valeria Falabella, Claudio Campagna, and John Croxall. Buenos Aires, Wildlife Conservation Society and BirdLife International, 2009. Web. Nov. 2011. http://atlas-marpatagonico.org/gis-database.html

10   A Place of One's Own

Samantha Senda-Cook and Danielle Endres

Decades ago, environmental advocates realized that when people develop a sense of place in the natural world through outdoor recreation, they are more likely to support pro-environmental policies (Marafiote). Indeed, John Muir states, "Thousands of tired, nerve-shaken, over-civilized people are beginning to find out that going to the mountains is going home; that wildness is a necessity; and the mountain parks and reservations are useful not only as foundations of timber and irrigating rivers, but as fountains of life" (48). Published in 1901, these words are still true today; people flock to wilderness places, looking for "escape" from civilization. The irony of outdoor recreation is that although engaging in this behavior is likely to increase one's sense of place and pro-environmental sentiment (e.g., Brooks, Wallace, and Williams; Buell; Cantrill; Cresswell; Dickinson; Ewert, Place, and Sibthorp; Kyle, Absher, and Graefe), it also has the potential to degrade the very places in which recreators seek to experience the natural world. In this chapter, we examine how this irony is revealed in the discourse of outdoor recreators. Specifically, we argue that this discourse encourages finding a place of one's own as key to an ideal recreation experience in the natural world.

We examine a set of textual fragments including popular outdoor recreation discourses in a magazine article and outdoor recreation catalogues, and interviews with outdoor recreators to understand the implications of seeking a place of one's own. Specifically, we argue that this rhetoric not only preserves the nature/culture divide and its problematic assumptions but also risks endangering the very areas that are set aside for preservation. In this chapter, we first explain the nature/culture divide and its correlation with outdoor recreation. Then, after delineating our methods, we illustrate how disparate discourses create a perception that ideal recreation happens in wild nature and alone or with a small group of known others. Finally, we explicate the consequences that this discourse produces in terms of the boundaries between nature and culture, and patterns of ownership.

OUTDOOR RECREATION IN A NATURE/CULTURE WORLD

The nature/culture divide is the idea that modern humans and civilization are separate from the natural world. Even though humans are natural beings that exist in the natural world, society reinforces the notion that humans are separate from, and sometimes better than, nature. Ideas about where and what constitutes nature are the products of centuries of rhetorically separating humans and nature. In articulating the birth of the nature/culture divide, Neil Evernden purports that with the creation of the word *nature*, we created a dualism. The "fragile division," as he calls it, between humans and nature must be maintained constantly or risk deteriorating. Jonathan Gray emphasizes the precariousness of the division. Embodying a character, he asks,

> What makes a scene authentically natural or wild? But we know it when we see it, right? It's like pornography that way, I guess. We've all probably felt the disgust of finding beer cans in a backcountry campsite. Or the surreal wonder of a hawk roosting in a crowded and noisy city. (209)

Gray highlights that nature is all around us and part of us, yet we are conditioned to understand nature as something outside of ourselves.

Within the nature/culture divide, an idealized version of nature as pristine, wilderness emerges. Yet, it is important to distinguish between wilderness and nature. Whereas nature is the material world and its processes, William Cronon defines wilderness as "a large tract of land distinguished above all else by the dominance of nonhuman nature and the relative absence of human influence" (ix). Wilderness areas are places that have not been significantly populated by or modified by modern humans, many of which are protected areas such as national parks. Peter van Wyck explains, "the perfect ecological space must be one that is absent of *modern* humans" and their accoutrements but not necessarily absent "ancient" humans and their "artifacts" (77). These "ancient" humans, the names of which sometimes refer to still living indigenous people, become part of ideal nature. Considering the nature/culture divide, the relative absence of modern human civilization in wilderness areas easily lends itself to classifying it as nature and not culture. In other words, whereas wilderness and urban spaces are both parts of the material world, the nature/culture divide constructs wilderness as nature and urban spaces as culture. Therefore, when humans seek to build a sense of place in nature, they tend to seek out wilderness areas, such as national or state parks. David Louter describes going to a national park: "We leave behind urban sprawl, the roadside blight of strip malls, and the patchwork of fields, clear-cuts, and other signatures of people at work in nature" (3). That people appreciate nature more when they come into contact with its idealized form is a cultural product not an inherent response (DeLuca; Sax).

The nature/culture divide is both constraining and enabling for environmentalists; we see them both reinforce and attack it. Proponents of wilderness preservation tend to reinforce the nature/culture divide in their suggestion that humans and potentially damaging human technologies should be excluded from some places so that those places may flourish. For example, Dave Foreman claims that even though people are part of nature, they should limit their presence in nature because we have "escaped the natural checks on our numbers," which has "allowed us to temporarily divorce ourselves from Nature" (404). This perception that nature is separate from humans positions it as something special or sacred. Although these positive impacts—preserving some areas and feeling a connection with nature—uphold the nature/culture dualism, it can also produce negative effects.

Kevin DeLuca argues that the idea of pristine wilderness is also a weakness for environmentalism. He states, "Opponents of environmentalism often argue against designating areas as wilderness because such areas are not absolutely pristine, or they will put a road in an area and then argue it cannot be wilderness" (643). As we entrench the divide between nature and culture, we feel more disconnected from nature because we cannot see that it is all around us as opposed to just in faraway and sometimes inaccessible places. Daniel Dustin, Kelly Bricker, and Keri Schwab explain that the trend of children staying indoors is detrimental to their health. Continuing to view nature as something separate hinders our "growth and development" (4). In fact, a sense of place and feeling of harmony with nature can be cultivated anywhere. One could have a profound experience with the natural world in a city park or on a farm. Challenging the separation between nature and culture encourages this perspective. One way to do this is through language. As Milstein et al. have done, using compound words such as "eco-culture" and "humanature" help change the way we think about nature and culture. Developing a sense of place through outdoor recreation is another way people suggest breaking down the nature culture divide (see, e.g., Dickinson; Spurlock).

We argue that outdoor recreation discourse sends the message that the best way to experience nature and develop a sense of place is in wilderness places. Although outdoor recreation can be defined as anything done outside from jogging in a city to hiking in a national park, we narrow our analysis to outdoor recreation in wilderness places set aside for preservation and recreation, such as national parks.[1] The key distinction we are making is a place-based distinction between outdoor recreation in places people perceive to be "nature" as opposed to places people perceive to be "culture" in line with the nature/culture divide. Outdoor recreation in these settings can including hiking, mountain bicycling, camping, and mountain climbing. The discourse associated with these activities, we contend, perpetuates a rhetoric of finding a place of one's own in which to have an "authentic" experience with nature, as conceived of in the nature/culture divide. However, as we will show, this rhetoric risks participants seeking more remote places to have their experiences. When people understand nature in opposition to culture, finding a place of one's own becomes challenging. People venture further afield to avoid other people because they want the ideal experience. The fragments of discourse we examine implicitly argue that crowds deny the opportunity to create a place of one's own, which leads to visitors seeking other, less populated areas in which to recreate and feel connected to nature. As Gray intimates, "I experience a kind of disappointment when I realize that there is probably no place on this planet that I can go to get away from the presence of other humans" (209). The danger is that the desire to be in more remote wilderness places can actually damage them.

METHODS

To examine how some forms of outdoor recreation discourse might encourage finding a place of one's own, we selected three texts: interviews with outdoor recreators,[2] a sample of outdoor recreation catalogues, and a *National Geographic* article on the best hikes. We do not conceive of the interviews as representative of all recreators at all national parks. However, the interview transcripts constitute a case study that reveals insights about the assumptions and approaches recreators bring with them. We collected twenty-five catalogues from a variety of sources with an eye toward including local and international brands and multiple years (see Appendix). We selected a hiking-focused article from a well-known, information-driven source. A *National Geographic* article titled "World's Best Hikes," in which Peter Potterfield, the author, picks the best hikes from his 2005 book on classic hikes, exemplifies the genre of outdoor recreation guides. Although not representative of the entirety of outdoor recreation discourse, these three texts represent how discourse circulates through mainstream sources compared to discourse produced by recreators themselves. This enabled us to see how the conceptualizations of being away from unknown others in nature aligned with, contradicted, or complemented one another across these texts. These artifacts came together to provide a picture of a dominant discourse in outdoor recreation, particularly as related to hiking. We analyzed them using a generative method of criticism, conducting initial analysis for emergent themes, developing a framework, and then subsequent analysis to gather specific examples.

A PLACE OF ONE'S OWN

The outdoor recreation discourses we examined demonstrate how fragmented texts create an argument for the desirability of finding a place of their own by venturing to remote, crowd-free, wilderness places. Our initial thought was that that most recreators desire solitude. However, our three texts complicate this interpretation. Although being alone was important, being away from unknown others was more important. In other words, retreating to *a place of one's own* means being able to decide who can and cannot be in the area. Just as one would not expect strangers to sit down on the living room couch, having a place of one's own implies that the most desirable way to experience nature is either alone or with a small group of friends or family. To get to such places, discourses suggest that recreators venture to remote places.

Wild Nature

Our analysis indicates that outdoor recreation in wilderness places is set up as the most desirable form of experiencing "nature." As outdoor recreation becomes more popular as a way to connect with nature, people seek more remote places to recreate. In other words, people feel a greater connection with a "natural" place when the place feels remote and uncivilized, thus entrenching the nature/culture divide. This perception constructs wilderness areas, areas away from civilization, as the best places to recreate.

All of our artifacts emphasized the desirability of wilderness through explicit words or images. Words supported the idea that recreation away from human culture was the best kind. Potterfield refers to "wild" places, "seemingly endless beaches of blinding white sands and surreal rock formations," "remote" locations, and a forest "so dense it seriously complicates navigation." In an interview, one woman described a climbing trip she took as her best outdoor recreation experience. She said, "It was a nice way, for 30 days, to get away from everything man-made." Many respondents stated that "escape," "to be away from the city mostly," and "to be more closely connected to the natural world and away from the busy, civilized world" were reasons they recreate outdoors. Even though they conceptualize outdoor recreation in broad strokes, their best experiences and the reasons for doing it at all strongly correlate to escaping the confines of culture.

In terms of images, the covers of catalogues never stated that people were in wilderness areas, but implied it through a lack of structures, paths, and people. They were never depicted in places crowded with people like cities but going even further, there was a noticeable lack of culture in the photographs as well. The cover photos did not show people in cabins or cafes but did show them outside, in open spaces. Even more significant were the photos of people on paths, evidence of human (and sometimes animal) use. The paths on which people walked, when they were there at all, were the only ones visible, implying the remote location. The images included in "World's Best Hikes" reinforced these norms. Although people were in almost every photograph, typically there was only one or two present in a seemingly natural place. And in only two exceptions did the viewer see evidence of cultures—towns in the background or many people walking.

These descriptions of getting into the wild or escaping culture point to one implied benefit of being away from people and culture: challenge. In finding a place of one's own, many recreators want to be challenged by being away from comforts and immediate help. They want to experience the "awe-inspiring," "amazing," "challenging," "wild," and "off-the-beaten-path" places as well. They find these places by reading magazines, trail books, and "guide books or enthusiast websites." They are willing to venture quite far to achieve the sense of newness and novelty. In researching places beforehand, participants can find out important information. The fact that the covers of the catalogues, the *National Geographic* article, and participants' answers align is no coincidence. It is impossible to say if the magazine articles make people search for more remote places or if people's desires drive the choices of catalogue manufacturers. Regardless, they unite to present a message about where to go to feel connected to nature and what constitutes and spoils nature.

Avoiding Crowds

Contemporary discourse about outdoor recreation supports the idea that connecting with nature happens in solitude and with known others. Explicit mentions of "avoiding crowds" or "being alone" illustrate how people construct a sense of place in nature. For example, in Potterfield's first entry in "World's Best Hikes," he describes a hike in Sweden: "The vibe here is 'far north,' with palpable emptiness and low-angled light that stirs the soul." "Emptiness" refers to an absence of humans and their technologies. Nature cannot be empty because it is full of all sorts of things like animals, trees, snow, and sky. This theme of seeking out emptiness and "hav[ing] it to yourself" surfaces in nine of the fifteen hikes described in the *National Geographic* article. In reference to the Grand Canyon, he advises, "Everybody does this hike in September to October or April to May, so go in March or November for a more contemplative experience." Potterfield recommends times to go and less popular routes that will allow the reader to avoid crowds: "May to September for drier weather; April or October for more solitude." Potterfield never explicitly explains that nature is supposed to be enjoyed alone, but he strongly implies it.

Our interviews confirm this finding with people reporting again and again that they try to get away from busy areas when they recreate outdoors. One person stated.

> I prefer places that are close and not crowded. This is usually a trade off. I will occasionally choose a trail/lake/mountain that I know are of lower quality (i.e. less maintained, not as nice) if it means I have a lower chance of running into a lot of people.

This idea of getting away from people was a common theme in the interviews. However, we also found that people wanted to develop relationships with known others when they recreated. People reported wanting to "be social," develop "fellowship with friend," "spend time with friends/partner," and simply "be with my girlfriend" as reasons they recreated. Notably what people are not seeking are new friendships. Nobody said that they recreated to meet people, encounter people who are different from a normal social circle, or feel the excitement of being with many other people.

In explaining where they like to recreate, people expressed that a lack of other people was desirable. Even those who saw themselves as outside the outdoor recreation subculture confirmed this finding. One self-identified "sight-seer" said, "I don't want to go there because at the height of the season, on a weekend, it's going to be full of tourists." One couple reported that they "never camp in campgrounds" because they want to experience being alone together. One of them stated that they were going to hike more trails in the frontcountry of Zion National Park, but "frankly we were ready to get away from the people. National parks are really oppressive to us." They said explicitly,

> We're disappointed by the national park crowd density. I mean, we know it's happening; we know it's going to be there, in a sense. But you go there and you kind of hope it's going to be better than it is. And no, it's not better than we thought. It's as bad as what we thought.

This sentiment pervaded the interviews among a wide range of visitors. One who was camping in the administrative campground in Zion National Park explained that it was

> nice because of the mere fact that I don't have to deal with all these tourists. You might not always like it, but you are going to have people—no matter where you go—that are going to be trying to get into the country, trying to get into the environment that you are in.

A local resident told me where he liked to recreate best: "I usually go up on the east side of the park and just find a place and just wander around back there, try and get away from the people."

The images on the catalogue covers and photographs that accompany the Potterfield article also communicate the message of a place of one's own by showing only a small group recreating instead of showing no one at all or hoards of people. Most (twenty-two) of the twenty-five catalogue covers we analyzed, and most (eleven) of the fifteen "World's Best Hikes" photographs featured between one and four people engaged in hiking or other outdoor activity in nonurban places. It is important that most of these images do not show nature in its ideal form (i.e., completely absent of people) but rather in the ideal form for outdoor recreation (i.e., with a small group of known others). The small groups imply that these people recreate together but not with unknown others, which communicates a feeling of ownership. The text and images indicate that by choosing the least crowded time of year to hike a certain trail and choosing the people with whom to hike, recreators exercise some control over how many unknown others they encounter. Although the people are not completely alone, they have the opportunity to be. Or if they want to be with other people, they can invite friends and family to join them. In most public places, people do not have this kind of control. At a crowded restaurant or movie theater, people must sit where they can regardless of who is near. But at home, a place of ownership, people can choose who joins them. The photographs communicate this possibility in nature, and make it desirable, by showing almost only small groups of recreators. This discourse cultivates the idea that people can find a place of one's own in wild nature and away from unknown others.

IMPLICATIONS

Our analysis reveals a discourse within nonurban outdoor recreation that stresses finding a place of one's own to truly experience nature. Overwhelmingly, this discourse highlights the desirability of engaging in outdoor recreation either alone or with a small group of known others in a wilderness area away from the signs of human culture. Although this discourse does support the idea of developing sense of place and experiencing the natural world, we argue that it does so in problematic ways.

First, the emphasis on being away from other people reifies the nature/culture divide. By challenging this rhetorical construction of a place of one's own, we join other scholars who argue that we can cultivate more ecologically sound practices by breaking down the nature/culture divide and pointing out the similarities between humans and nature (e.g., DeLuca; Milstein et al.; Sowards). The desire to experience more and more remote places enforces the idea that nature can only be found in places that are not crowded with humans and their technologies. However, with the constant drive of outdoor recreators seeking new less-crowded places, this may unintentionally reduce the amount of land considered to be wilderness. In other words, the new peopleless places can only stay peopleless for so long, thus motivating people to seek out new places. This can lead to the degradation of the environment in two ways. In one way, this reification of "real" nature as peopleless and wild can lead to the destruction of other places that are not perceived to be natural such as cities or even crowded parts of national parks. In another way, this desire to see "nature" in remote wilderness places can lead to the actual degradation of those places. People's presence, even if not directly destructive, can have untold impacts on the ecosystem. And the more people who come to these remote places, the more risk of destruction.

We recognize the potential contradiction in our argument that we need to protect these remote wilderness areas from the destruction of people and our deconstruction of the nature/culture dualism. However, we do not see it as a contradiction. Certain places need to be protected from human use and degradation—in both wilderness and urban areas. Val Plumwood illustrates the oppositional consequences of the nature/culture divide, how we at once need the distinction between nature and culture and how it harms the environmental case. She states,

> Without some distinction between nature and culture, or between humans and nature, it becomes very difficult to present any defense against the total humanization of the world, or to achieve the recognition of the presence and labor of nature which must be a major goal of any thoroughgoing environmental movement. For that we need sensitivity to the interplay of self and other, and to the interweaving and interdependence of nature/culture narratives in the land. But we need not and should not construct the distinction as a binary opposition, as the Western dualism of nature and culture has done. (676)

The argument that some places need to prohibit human use does not necessarily entrench the idea that nature and culture are separate. Rather it suggests that we need to limit human impact on the environment—not because it is pristine and without humans but because places need to recover from human damage.

The second implication of our analysis is that finding a place of one's own may be less about developing a sense of place and more about human competition and ownership. In other words, the aspect of finding a place of one's own that involves challenge, exploration, and adventure emphasizes being able to control how one experiences a place, which connects to the idea of conquering nature. One respondent said that (s)he hiked off trail "because trails are too confining. I'd rather have an original experience with nature—not shared by anyone else." Relph argues, "A sense of place that stresses uniqueness to the virtual exclusion of a recognition of shared qualities is an ugly and violent thing. It is indeed a poisoned sense of place" (223). In this sense, the stressing of finding a unique and special place may encourage a sense of ownership of nature.

In the end, we understand the power of peopleless experiences in nature toward developing sense of place and an associated environmental ethic. We have both had these types of experiences. John Muir, Henry David Thoreau, Dave Foreman, Terry Tempest Williams, and Rebecca Solnit describe these experiences. Rather, our critique focuses on how outdoor recreation discourses of finding a place of one's own, although seemingly pro-environmental, may actually be harmful to the natural world through encroachment, feelings of ownership, and, worse, material damage to the environment. At risk of reinforcing the nature/culture divide, we counter the outdoor recreation discourse of a place of one's own with a call to let animals and plants have a place of their own, just sometimes.

APPENDIX: LIST OF CATALOGUES

Canada Goose, Canada, 2008. International product manufacturing.
Icebreaker, New Zealand, 2006–2007. International product manufacturing.
Icebreaker, New Zealand, 2006. International product manufacturing.
Kirkham's, U.S.A., 2007. Local store.
Kirkham's, U.S.A., 2008. Local store.
Kirkham's, U.S.A., 2008–2009. Local store.
Kirkham's, U.S.A., 2009. Local store.
Patagonia, U.S.A., 2006–2007. International product manufacturing and stores.
Patagonia, U.S.A., 2009. International product manufacturing and stores.
R.E.I., U.S.A., 1972–1973. International product manufacturing and stores.
R.E.I., U.S.A., 1973. International product manufacturing and stores.
R.E.I., U.S.A., April, 2007. International product manufacturing and stores.
R.E.I., U.S.A., May, 2007. International product manufacturing and stores.
R.E.I., U.S.A., June, 2007. International product manufacturing and stores.
R.E.I., U.S.A., "Fall Preview," September, 2007. International product manufacturing and stores.
R.E.I., U.S.A., "Labor Day Sale," September, 2007. International product manufacturing and stores.
R.E.I., U.S.A., October, 2007. International product manufacturing and stores.
R.E.I., U.S.A., "Winter Sale," November, 2007. International product manufacturing and stores.
R.E.I., U.S.A., September, 2008. International product manufacturing and stores.
R.E.I., U.S.A., October, 2008. International product manufacturing and stores.
R.E.I., U.S.A., "Warm-Up-For-Winter Sale," December, 2008. International product manufacturing and stores.
R.E.I., U.S.A., "Celebrate It," December, 2008. International product manufacturing and stores.
R.E.I., U.S.A., May, 2009. International product manufacturing and stores.
Snow Peak, Japan, 2008. International product manufacturing.
Wasatch Touring, U.S.A., 2008. Local store.

NOTES

1. Different degrees of wilderness comprise national parks. Frontcountry areas include some of the following: roads, visitor centers, campgrounds or hotels, cafeterias, and gift shops. Backcountry areas usually require hiking for access and have fewer (if any) facilities.
2. A short (ten-question) online survey contained seventy responses and produced forty-three pages of transcripts. Twenty in-person interviews gathered at Zion National Park produced over 230 pages of transcripts. Both are available on request. One of us conducted the interviews at Zion National Park over the period of a month during May and June of 2009 in different parts (e.g., backcountry, on trails, in campgrounds, in the town outside the park, in non-public areas of the park) of the park to gain access to a variety of recreators. Participants were selected more or less randomly. We gained IRB exemption status for both sets of interviews.

WORKS CITED

Brooks, Jeffrey, George Wallace, and Daniel Williams. "Place as Relationship Partner: An Alternative Metaphor for Understanding the Quality of Visitor Experience in a Backcountry Setting." *Leisure Sciences* 28.4 (2006): 331–349. Web. 1 Sept. 2011.

Buell, Lawrence. "The Place of Place." *Writing for an Endangered World.* Cambridge, MA: Belknap-Harvard UP, 2001. 55–83. Print.

Cantrill, James. "The Environmental Self and a Sense of Place: Communication Foundations for Regional Ecosystem Management." *Journal of Applied Communication Research* 26 (1998): 301–318. Web. 14 Sept. 2011.

Cresswell, Tim. *Place: A Short Introduction.* Annotated edition. Malden, MA: Blackwell, 2004. Print.

Cronon, William. Foreword. *Windshield Wilderness: Cars, Roads, and Nature in Washington's National Parks.* David Louter. Seattle: U of Washington P, 2006. ix–xiii. Print.

DeLuca, Kevin. "Trains in the Wilderness: The Corporate Roots of Environmentalism." *Rhetoric and Public Affairs* 4.4 (2001): 642–643. *Communication and Mass Media Complete.* Web. 1 Sept. 2011.

Dickinson, Elizabeth. "Displaced in Nature: The Cultural Production of (Non-) Place in Place-Based Forest Conservation Pedagogy." *Environmental Communication: A Journal of Nature and Culture* 5 (2011): 300–319. Web. 14 Sept. 2011.

Dustin, Daniel, Kelly Bricker, and Keri Schwab. "People and Nature: Toward an Ecological Model of Health Promotion." *Leisure Sciences* 32 (2010): 3–14. Web. 1 Sept. 2011.

Ewert, Alan, Greg Place, and J. I. M. Sibthorp. "Early-Life Outdoor Experiences and an Individual's Environmental Attitudes." *Leisure Sciences* 27 (2005): 225–239. Web. 1 Sept. 2011.

Evernden, Neil. *The Social Creation of Nature.* Baltimore: Johns Hopkins UP, 1992. Print.

Foreman, Dave. "Wilderness Areas for Real." *The Great New Wilderness Debate.* Ed. J. Baird Callicott and Michael P. Nelson. Athens: U of Georgia P, 1998. 397–407. Print.

Gray, Jonathan. "Trail Mix: A Sojourn on the Muddy Divide between Nature and Culture." *Text and Performance Quarterly* 30.2 (2010): 201–219. *Communication and Mass Media Complete.* Web. 1 Sept. 2011.

Kyle, Gerard, James Absher, and Alan Graefe. "The Moderating Role of Place Attachment on the Relationship Between Attitudes Toward Fees and Spending Preferences." *Leisure Sciences* 25 (2003): 33–50. Web. 1 Sept. 2011.

Louter, David. *Windshield Wilderness: Cars, Roads, and Nature in Washington's National Parks.* Seattle: U of Washington P, 2006. Print.

Marafiote, Tracy. "The American Dream: Technology, Tourism, and the Transformation of Wilderness." *Environmental Communication* 2.2 (2008): 154–172. Print.

Milstein, Tema, Claudia Anguiano, Jennifer Sandoval, Yea-Wen Chen, and Elizabeth Dickinson. "Communicating a 'New' Environmental Vernacular: A Sense of Relations-in-Place." *Communication Monographs* 78.4 (2011): 486–510. Print.

Muir, John. "Selections from *Our National Parks.*" *The Great New Wilderness Debate.* Ed. J. Baird Callicott and Michael P. Nelson. Athens: U of Georgia P, 1998. 48–62. Print.

Plumwood, Val. "Wilderness Skepticism and Dualism." *The Great New Wilderness Debate.* Ed. J. Baird Callicott and Michael P. Nelson. Athens: U of Georgia P, 1998. 652–690. Print.

Potterfield, Peter. "World's Best Hikes." *National Geographic*, n.d. Web. 4 Oct. 2011.

Relph, Edward. "Sense of Place." *Ten Geographic Ideas that Changed the World.* Ed. Susan Hanson. New Brunswick, NJ: Rutgers UP, 1997. 205–226. Print.

Sax, Joseph. *Mountains Without Handrails: Reflections on the National Parks.* Ann Arbor: U of Michigan P, 1980. Print.

Sowards, Stacey. "Identification through Orangutans: Destabilizing the Nature/Culture Dualism." *Ethics & the Environment* 11.2 (2006): 45–61. *GreenFILE.* Web. 1 Sept. 2011.

Spurlock, Cindy M. "Performing and Sustaining (Agri)Culture and Place: The Cultivation of Environmental Subjectivity on the Piedmont Farm Tour." *Text & Performance Quarterly* 29.1 (2009): 5–21. Print.

Van Wyck, Peter. *Primitives in the Wilderness: Deep Ecology and the Missing Human Subject.* Albany: State U of New York P, 1997. Print.

11 Local Flaneury
Losing and Finding One's Place

Jaqueline McLeod Rogers

To lose oneself in the city—as one loses oneself in a forest—that calls for quite a different schooling. (Walter Benjamin, in Kingwell, 242)

Probably more than any other cohort, college students—particularly those who are younger, just entering adulthood—are deeply invested in wrestling with the question of where to live. For many, staying in place is a habit they are eager to break. But not everyone seeks new ground. Another large group, familiar with the comforts of home, resist going away and making changes. Asking students to pay attention to local place by adopting flaneury (from the French, *flâneur*) as a form of inquiry can help them to see what is really happening in the material, social, and linguistic environment. Rather than seeing the local world as full of endless repetitions—either boring or comforting, but seldom out of the ordinary—student-flaneurs can enlarge their awareness of environmental elements and ecology and develop a sense of agency in relation both to how they perceive and move in the world. The practice of flaneury—strolling, or slow walking, through the urban landscape—is a way to reframe one's place in urban ecology. Such thinking about one's position in place allows for more informed decisions about staying or going.

Historically, flaneury is urban work, suited to knowing the built environment and examining human dreams and social relations in this world. The material city is mostly built environment and ever-increasing technological circus. To walk through the city is to give up relying on the usual transportation routes and technologies, and to think about the place of human bodies in this maze. Rather than observing changing street scenes from a stationary position, the flaneur moves within the changing scene and establishes a sense of self as being within the urban ecology. Moreover, to study place by moving through it on foot is to adopt a natural rhythm or pace, rather than continuing to move to the pace of the technologized city.

Theorists speculate that above others, the purposes of human safety and communication were served by the founding of cities (Araujo 200). We feel safe in the city for we are inside—inside its boundaries, inside our homes, inside our skins—yet ages of city living have shown that there are dangers, too, in being overly interior—in being locked inside structures or our bodies, so that we are alienated or isolated. Local flaneury is about involvement and movement, without going to the extreme of immersion or absorption. It is about examining boundaries, by moving through places where one feels at once both an insider and outsider. By walking down an unfamiliar city street, the flaneur secures a new sense of connection to place, but a connection that is passing rather than intimate, and might be conceptualized along these lines: "This is my city, but this is not my neighborhood; this is not my neighborhood, but I have walked through it and now know something of it." Rather than driving mindlessly through city streets or dismissing a block as a place "I never go," the flaneur initiates a respectful connection to place, which is now known as both familiar and unfamiliar, in acknowledgment of the layered complexity of place.

The city also affords communication technologies—both visible, like roads and sidewalks, and invisible, like Wi-Fi connections—that we rely on without thought, unless street construction forces a change of route, or the Internet is interrupted. Flaneury encourages practitioners to be more conscious of how roads, street signs, and lights map concerted effort to exercise some options to determine our movement, perhaps even to resist the flow of the crowd. Of course it must be conceded that cities are heavily designed and regulated, yet flaneurism allows practitioners some space to push back, to bend some of the straight lines—to find shortcuts, alleys and places to pause.

Apart from becoming more conscious of one's relation to place—of inhabiting a space that is both familiar and strange, closed and open—the moving flaneur is also positioned to think about the work of words in the social world, both in the signs that are everywhere and in spoken language. Examined close up, the linguistic landscape does more than communicate the messages of its producers, for as well as telling about the assertion of power and control, it also tells of competition and creativity. There is the language of official communication, but often the street talks back, with creative transgressions that offer alternative commentary or solutions. Rather than shutting out signs and words as another taken-for-granted regulatory event, part of flaneury is to actively listen to the local language and read the signs along the way, paying special attention to the question of how words are changed in context, so that as language theorist Alastair Pennycook reminds us "what appears to be the same may in fact also be already different" (*Language* 49).

The ground I want to explore in this chapter is mostly urban, a place where students can think about the differences between humans and the tools and technologies we have built to communicate, to make us feel safe and move us around. A city is not an interesting place if I shut my eyes to how it is built or if I feel that I have no control over how it runs. Flaneury as a form of inquiry provides a way to abandon the blinkers of routine to perceive the rich and networked urban ecology.

HISTORY OF FLANEURISM: WHAT'S A FLANEUR TO DO?

Many of us recognize the term "flaneury" from studies of nineteenth-century literature as referring to the dandy's way of watching the urban parade of strangers and commodities. The artist as dandy attempted to observe and then depict lives in public places, often by standing apart from his subjects, with colonizing gaze and barely concealed attitude of slumming. The dandy often saw vigor and beauty in those he described, yet was observer rather than part of the group and scene. Flaneury can be understood as the fieldwork of nineteenth-century dandyism, allowing the urban artist to pick up details of the social world. The flaneur was not nostalgic for nature or solitary reflection, but excited to document, critique and lyricize what seemed the ceaseless energy and novelty of the urban social parade. For wealthy or bourgeois society at the time, there was a general belief in the material happiness that was going to be brought about by capitalism and comodification, and the cities sparkled with the gas lights and glass of the arcades built to celebrate sales and opulent consumption. The flaneur acknowledged urban beauty and hope, without entirely sharing the attitude of optimism about urban comfort and progress, capturing in art the stinging forces of opposition in the faces of the poor and marginalized.

Most often cited as giving definitive voice to this early flaneury is the work of French poet Charles Baudelaire (1821–1867). As much as he is intoxicated by the city, the artist as flaneur remains an outsider figure. In this first poem from *Tableaux Parisiens*, for example, he surveys the city from a bird's eye view, captivated by the bright images yet viewing the spectacle from above and apart:

Chin on my two hands, from my mansarded eyrie,
I shall see the workshop with its song and chatter,
Chimneys, spires, those masts of the city,
And the great skies making us dream of eternity. (Donald 3)

Baudelaire's prose poem/short story "The Eyes of the Poor" addresses themes of attraction and detachment, depicting the protagonist's fascination with a poor family who has no place in the opulent cafes of Paris. Although the protagonist responds with empathy beyond that displayed by other members of his wealthy class—his own fiancée regards the poor family with judgment and disdain—class difference prevents him from knowing what they are actually experiencing. He is captivated and provides a window onto lives otherwise unseen, yet he remains an outsider figure.

Walter Benjamin revived interest in the flaneur in the 1930s, taking up the role in a world that no longer dreamed of capitalist industry bringing happiness. Urban life had grown old, devolved into commodification, mass reproductions, and social strife. The arcades were no longer glittering retail and restaurant palaces, but grimy, broken and unpopulated. Yet by strolling through the old arcades, Benjamin found a residual beauty, an archive of lost dreams. It was even possible to see an element of freshness and timelessness in items from the past, what he called an "aura" of a time when things were seen as new and even original.

Whereas Benjamin's prewar European city was one of endless reproduction, deterioration, and darkness, his flaneury is understood as an act of hope, for he turned to it as a way of seeing the aura and hearing echoes of a more energetic past. Sven Birkerts describes Benjamin's flaneur as a questing figure who "because he has abandoned the hierarchy of accepted values and modes of perception, is the medium through which the original connections and relations have a chance to disclose themselves" (170). Yet this figure is seldom able to recover a sense of aura or authenticity, because "the harmony between man and his world had been all but irreparably violated" (176). Whereas others of his Frankfurt School, Horkheimer and Adorno, were critical of comodification and capitalism, Benjamin was haunted by the promise and beauty once posed by the world of things, and sought for hints or fragments of this aura in the fading dreamscape.

In the 1990s, with a rising interest in spatial theorizing, interest in flaneury revived again. As always, it offered a way of observing the urban landscape. Yet spatial theory itself offered a way to understand the flaneur as a figure associated with vision and change. James Donald describes how, for spatial theorists fascinated by the interaction between inner and outer spaces, "the key to understanding the dynamic . . . between social space and mental space is the *transition*" (13). Flaneury is a way of making such transitions visible, allowing the individual to be aware of the way one's mind and body shape and change the environment as well as how the environment shapes where one walk and how one thinks. The flaneur is a "border" or perhaps liminal figure. He is not immersed in the social activity of place, but by passing through it he is always on the border, both entering and leaving.

Mark Kingwell is another contemporary spatial philosopher who has explored "the threshold" as a powerful transitional position. Indirectly, he connects this perspective with the movement of flaneury by suggesting that being rooted and stationary restricts what one can see or imagine. He contends that people who are steadfast in inhabiting places fail to understand their composition as well as possibilities and opportunities for change. The flaneur, not locked into a rooted perspective, is associated with movement and vision: "The function of the threshold, therefore, is . . . to *separate* and thus *to be crossed* . . . Drawing the limit line inspires the desire, or demand, that the line be crossed. Once established, boundaries 'ask for' breaching— traditionally a task for heroes" (158). The flaneur continuously on the move is continuously crossing thresholds rather than respecting impenetrable boundaries and separations. Yet it also true that liberating as it is to be able to break through boundaries, we rely for day-to-day order on boundaries to separate inside from outside, private from public space. Flaneury is not so much about breaking down barriers as widening margins, for the flaneur is not imposing him or herself, not moving in to occupy a new space so much as moving by and through it. In the performance of local flaneury, there are elements of both dwelling and leaving, for the flaneur is both insider and outsider as he/she moves through places that are connected to home and known places, but that as layered palimpsests of complexity are resistant to intimacy.

Within cultural geography, flaneury is often connected to borderland "thirdspace," a concept popularized by Henri Lefebvre's triadic model of thinking in *The Production of Space*. Rather than asking us to think in dialectical fashion, Lefebvre proposed three interactive spaces, none dominant or fixed. In *Geographies of Writing*, Nedra Reynolds provides a solid introduction to Lefebvre's three key terms. "Conceived" space refers to material space and physical places, and the ways we measure and plan to organize this space. "Perceived" space (also called "spatial practice") has to do with the material and social world we know through sensory perception, which takes in habitual social practices that produce routine coherence. Finally, "lived" space has to do with how we actually use and negotiate space, the space where we are able to make unusual moves and engage transgress against habit.

According to Edward Soja, "lived space" is the dimension closest to third space, "the space of radical openness, the space of social struggle": "With its foregrounding of relations of dominance, subordination, and resistance; its subliminal mystery and limited knowability; its radical openness and teeming imagery, this third space of Lefebvre [what he called *spaces of representation* or *lived space*] closely approximates what I am defining as thirdspace" (Soja 68). Lived space, the space of social and physical movement where individuals can imagine and enact change, is connected to flaneury, which is about moving through the social and physical worlds to see new possibilities. In a biographical sketch of Lefebvre, Soja offers another link between thirdspace perception and flaneury. He tells us that throughout his life, Lefebvre traveled between his Paris apartment and his birthplace in an isolated village in the Pyrenees. Soja argues that it is neither the urban nor the rural location that defines Lefebvre as a thinker or that provided him with a definitive sense of homeplace; instead it is his ability to move between and embrace both places that explains Lefebvre's creative perspective thusly:

> The "prodigiously stimulating" dialectics of center-periphery and conceived-lived were concretely re-enacted every year for decades in Lefebvre's oscillations between Paris and his home in Navarrenx . . . a site filled with passions and memories of radical regional resistance to centralized state power. From these regenerative movements, practically all of Lefebvre's achievements and inspirations can be drawn. (31)

If the flaneur's marginal status frees him to think more creatively—clearing a "radical openness" to allow for imagining thirdspace options—it also has political implications. More than simply standing to gain an improved personal understanding of place, or sharing this insight through art and creativity, he or she is also positioned to recognize and enact civic change. To adopt urban analyst James Donald's terms, we urban dwellers have the capacity to be both "the good citizen" and "the man about town" (this last position being one he uses as closely synonymous with "the flaneur"), and we need both dimensions to be fully alive in the city. The good citizen works to establish an ideal city and tries to improve public places; the man about town, without civic agenda, is free to see firsthand all that actually exists in the city—back alleys and dark corners, as well as main street public spaces. Donald argues that it is dangerous when we allow these ways of seeing to become divided or work against each other—we can become like Jekyll and Hyde, working for goodness and perfection by day yet pursuing our appetite for self satisfaction by night (118). For Donald, the impulse to build a better and safer world can work in tandem with the impulse see the world fully and creatively. We need law and order, yes, but also to be able to seek self-satisfaction, personal insight, and growth. Donald is concerned that the current intellectual climate favors social justice agendas and cultural critique and he cautions that "the good citizens and zealous republicans should leave us flawed men and women about town some space" (119).

Whereas the flaneur figure of the nineteenth century was more strongly associated with apolitical art and creativity than with being public minded and socially conscientious, the contemporary flaneur is a more accretive figure, with a moving foot in both camps. Although he can cultivate personal insight, by doing this he can also imagine new and better "thirdspace" alternatives. Yet such alternative policies that are the product of flaneury need to be understood as provisional and dynamic rather than programmatic, as mobile as the form of inquiry by which they were devised.

Thus, practicing contemporary flaneury involves cultivating both creative vision and political awareness. Moreover, practiced as a form of inquiry, this action is self-conscious, so the flaneur is aware of making connections amongst seeing, thinking, and doing. This is a powerful epistemic.

BREAKING HABITS, GOING PLACES

The practice of flaneury takes aim at lazy habit. Walking through areas we know with eyes open we see new things, as the way is defamiliarized. Additionally, practicing flaneury can take us to unknown, even hidden places. Rather than staying with known routes, the walker frequently finds new paths, and thus expands his or her conception of what the city holds.

In an article called "Inhabiting," British writer Geoff Dyer makes the case that many of our habits are spatially rooted, and we move almost automatically through the spaces we know. He suggests we turn off thought and perception, and are guided by a sort of embodied automatic pilot. He reminds us of the distinction Nietzsche made between the dead weight of "enduring habits" and the leaven of "brief habits" (160)—those which save us from floating aimlessly without any sense of pattern or plan—before telling of his own efforts to lift anchor from old habits to negotiate more freedom of movement—to expand his repertoire and adhere with less devotion to patterns of action. Dyer recognizes his fondness for habit leaves him living on Effra Road in London for years, himself virtually changeless as buildings and social life change around him. When he eventually moves to New York, he attempts to recreate and repeat the same habits he has always observed, and thus he searches for a coffee shop that serves pastries like those he has always had during his 11 a.m. break, hoping to reinstate "the unvarying routine I crave and need" (162). In New York, he discovers a donut that is better than any other he has had before, and makes getting this treat the highlight of each day. Despite this unexpected positive outcome, Dyer remains anxious about being ruled by compulsion or obsession. He observes that most often change of place means nothing if we are psychologically conservative, so that "when we talk about life we might just as well talk about trudging up and down Effra Road, irrespective of where we are in the world, for we are always trying to recreate our particular idea of a city in whichever actual city we happen to find ourselves" (168).

Ultimately, Dyer's narrative ends on a more positive example of engaging in a "brief habit." No longer alone, but casting himself as a tourist with his wife as companion, he walks the streets of Tokyo, initially worrying about the repetitions wrought by global culture. There is a Starbucks or MacDonald's in every port. Yet by chance he finds evidence that there may be richness and rewards in this repetition when he discovers a coffee shop with the exact brand of donut he had come to count on in New York. This discovery leads him to a layered reflection on sameness and difference, on self and other. Salient to our argument here is his recognition that people are changed by moving, that places have commonalities and differences, and that some of the changes and differences are positive innovations, symbolized in his story by the appearance of the perfect New York donut in Japan, a donut that cannot be had on Effra Street in London. This is his insight about Tokyo, a "strange, strangely familiar city" (169):

> There were ancient temples and carp ponds, there were geishas and cherry blossoms, there were incredible feats of retail architecture and there was the sci-fi neon of Shinjuku, but at the back of my mind, in contrast to the untranslatable otherness of Japan, was the familiar prospect of Donut Plant donuts which, far from being reminders of home, were, for most of the year—the part of the year spent in London—site-specific New York treats, edible emblems of everything that New York was and London was not. (168)

By moving and breaking with dulling and enduring habit, he found space composed of similarities and differences. Although his journey is global and the flaneur stays local, there is overlap in the learning outcomes: things change and echo familiarity from city to city and block to block, yet if one pays attention—walks with "radical openness"—one has a chance of encountering a new object or configuration.

Recent science studying human memory and routine supports Dyer's observations about the neural depth of our human commitment to habit. Brain researchers have found we execute habits without thinking because patterns of habitual doing are stored in an old part of the brain that resists change. Popularizing brain research, journalist Charles Duhigg recently published *The Power of Habit*, an examination of the difficulties and possibilities of unseating routine. He describes how many of us resent being controlled by habit and feel self-loathing when we are unable to discontinue a routine, despite our best resolve. He points out that our susceptibility to habit leaves us vulnerable to marketing specialists who have studied ways to control and manipulate patterned behavior. He suggests that we are likeliest to succeed changing a habit if we substitute it with another like it, satisfying our craving by means of using a placebo. What I am suggesting as a way to break free of the hold of habitual seeing and doing is adopting a new more self-conscious way of seeing. Teaching students to practice flaneury is to teach them to adopt an attitude of iconoclasm that goes beyond a simple form of contrarianism, in being a more layered response. Rather than simply perform habit without thinking, they can think about what they are doing in social and material space, and from this go on to imagine how these worlds of action and things might be reconfigured in a new interactive constellation. If our body resists change by storing habits, then asking students to practice flaneury provides them with a stretching and loosening exercise, awakening perception and fueling mobility.

TRANSGRESSING TECHNOLOGIES: PLAYING IN THE TRAFFIC

Apart from helping to free us from psychological dependencies, first by making them visible, and then by removing them from easy access, the act of flaneury also breaks us out of our day-to-day reliance on technologies by forcing us to change channels. As early as the 1960s, Marshall McLuhan registered his concern about the tendency of most people to simply accept without questioning the quickening pace of the life-changing barrage of technology. He frequently called for people to abandon "voluntary somnambulism" and cultivate instead a more perceptive attitude toward the urban environment (3). In a late-career book called *The City as Classroom*, written with Kathryn Hutchon and Eric McLuhan, he urges students to set aside print texts to go outside and read the world—to leave the classroom for the street to see how many ways we depend on technologies to carry out everyday business. His book provides exercises to help students break through the ease of living in a taken-for-granted environment to begin to see, hear, even feel the physical and social world. His key point was that paying attention to everyday life conjures phenomenological insight.

There is a link from his methodology to flaneury. Walking, for example, frees us from modes of transportation that are so much a part of city apparatus that we seldom question their regulatory function. There is little question life is easier and safer because cars move us long distances over lighted roads, but to simply depend on these technologies is as dangerous as assuming the air we breathe is clean. Part of the business of the flaneur is to measure the pace of his or her movement against the faster paced movement of wheeled traffic, to understand urban pace and motion as an ecology rather than a dichotomy between the human body and technology. In her book *For Space*, Doreen Massey reminds us that the natural world itself is on the move, that even though we imagine places have permanence or fixity, the rocks of Skiddaw, like other geographical features, follow a "migratory" pattern—"are immigrant rocks, just passing through here, like my sister and me only rather more slowly, and changing all the while" (137). Flaneury grants movement, and puts us in position to be more aware of other motion in the world; it exercises and develops the spatial imagination.

Moving, walking, the flaneur need not follow a predetermined or even a publicly sanctioned route. His or her way is not determined by street technologies, but directed by personal agency, for the flaneur chooses route, direction and deviations, being above all open to chance and change. Connecting flaneury to the Situationist *derive*, or "drift," Mark O. Turner proposes the alternative self-explanatory term "zigzagging" to capture the importance of following one's nose and inclination rather than a map, being open to serendipitous paths and conveyances rather than always guided by prescribed route. Rather than following straight lines or the shortest route from A to B by zigzagging, the flaneur makes his or her own way, often going against the grain of other street traffic, often veering off regulated or authorized paths. One's direction is not dictated by street technology, but by desire.

Turner emphasizes the transgressive power of this "wandering" and suggests it can be a broad-based mode "of resistance against all those attempts to plan and order our lives" (307). He reminds us that de Certeau was another who theorized walking against the flow as a "tactic" to resist the institutions that regulate life, a tactic being

> [a] mobility that must accept the chance offerings of the moment, and seize on the wing the possibilities that offer themselves at any given moment. It must vigilantly make use of the cracks that particular conjunctions open in the surveillance of the proprietary powers. It poaches in them. It creates surprises in them. (qtd. in Turner 308)

Turner reminds us that the city already has hidden passages, dark alleys, and an underground, and that by walking we may find locations not on the map like these, and thus expand our notion of what and whom the city includes.

Rather than encouraging student flaneurs to go down dark and hidden streets into possible dangers, we can encourage the cultivation of a similarly expansive view by recommending that they change their angle of vision, by being aware of looking up and down as well as ahead. There is much to see walking slowly, but they should also be conscious of looking for and at what they usually do not see and of there being layers that remain unseen—the city that remains hidden, whether in the form of underground infrastructure or of radio and digital waves above. This flaneur—no longer lost in the crowd or pushed through the city in regimented automation—moves with agency through a city that, layered and mysterious, is worth such efforts at encounter and knowing. As they study material and social space, they need to keep in mind that some doors remain hidden or locked, that there is always more beyond the more they come to know.

Apart from roads and the machinery that regulate city traffic and transportation, language is another taken-for-granted tool of social organization. Our cities are full of signs that attract, direct, and regulate behavior. Because there are so many, with so much repetition amongst them, we tend to respond more or less automatically, internalizing language statements and then complying without thinking. According to Pierre Bourdieu, our worlds abound in "structuring structures" that "reproduce the perspective of those who create them for those who encounter them" (in Schaafsma and Vinz 82). By observing these structures rather than complying by rote, we can open up spaces and participate in producing generative meanings, which in turn opens new spaces for the body to act. As Bourdieu's term "habitus" conveys, we are both governed by prescribed discourse as well as able to define the terms of our compliance by "negotiat[ing] the world with other people" (83). Within this framework, flaneury can be understood as supporting an attitude of observation, rendering the flaneur assertive and less likely simply to do as told.

More poetically in *Invisible Cities*, Italo Calvino explores how signs and names serve as linguistic devices to routinize and restrict our perception of place. In the section discussing "Cities and Names," his reflection on "Aglaura" captures the way we come to know a city by name as one that has only a certain number of specified qualities, and how this leads us to overlook exceptions or challenges:

> At certain hours, in certain places along the street, you see opening before you the hint of something unmistakable, rare, perhaps magnificent; but everything previously said of Aglaura imprisons your words and obliges you to repeat rather than say. Therefore, the inhabitants still believe they live in an Aglaura which grows only with the name Aglaura and they do not notice the Aglaura that grows on the ground. (68)

Other bars of the prison house of language are revealed when we apply Calvino's example to a real city place and name. If I tell you that my city is called "Winnipeg," and if you have had a friend visit Winnipeg and tell you it is a forgotten frozen wasteland, then the name will spark a negative image for you. If I support your friend's view, the negative association you have made between name and place grows stronger. Were you to visit, chances are you would find images that support your predisposition, and tend to overlook anything "rare" or "magnificent"; if as an independent thinker you were to see rare magnificence, you would be likely to note this quality in terms of its being contrary to your preconceptions. As a long-time resident of Winnipeg, my preconceptions of what the city holds and means are even stronger, which leads me to be even more likely to resist recognizing moments of contradiction. Through practicing flaneury, students can peel away some of the barriers erected by language repetitions and routinized discourse, to gain fresh first-hand perspectives on the literal material place as well as access to the unfolding social world of discourse.

To attend in a broad sense to the linguistic landscape is to take stock of forms of communication we often take for granted, including written signs, visual rhetorics, and the speech and gestures we perform. Not only is it best to conceive of this landscape in broad discursive terms, but, according to Alastair Pennycook, it should also be understood as in the midst of dynamic unfolding—not so much perceived as thing already performed but in active production:

> Readers and writers are part of the fluid, urban semiotic space and produce meaning as they move, write, read, and travel . . . Landscapes are not mere backdrops on which texts and images are drawn but are spaces that are imagined and invented. The social effects, the illocutionary force, of urban texts are animated by the movement and interaction of city dwellers. ("Linguistic Landscape" 309–310)

Pennycook's perspective startles the flaneur from the restfulness of being a language voyeur and eavesdropper to assuming the role of more active participant-producer of meanings.

Elsewhere Pennycook theorizes the advantages of understanding the local character of language, a character that can only be understood if one engages in the heat of practice. In *Language as a Local Practice*, he makes the argument that when we turn to a specific local community, we often begin by looking and listening for patterns that form the basis of a grassroots literacy and generally for elements of disparity that point to how the local differs from the non-local. But, eventually we hear echoes of other places and times, so that what we find in the local is "the same-but-different": he says, "We need to look more closely at mimesis and repetition if we are to understand how language and indeed human freedom work . . . [and] to undercut those dichotomous assumptions about sameness and difference, creativity and conformity, repetition and originality . . . Language diversity/creativity is about sameness that is also difference" (51). What

Pennycook is saying is that we put our finger on creative energy when we identify how sameness is made different—how language, for example, has a unique local life.

The photo on the cover of Pennycook's book—a photograph of the contemporary art statue, "The Reverend on Ice"—helps to illustrate the creative energy of sameness and difference in local and contemporary life. The photo is of a sculpture created in 2005 by Yinka Shonibare, a black artist born in Britain but brought up in Nigeria. The sculpture is one of a series of headless and thus de-racialized figures. The piece is based on a painting from the 1700s by British painter Henry Raeburn called "The Reverend Robert Walker Skating on Duddingston Loch." Shonibare's sculpted figure wears fabric that the artist found advertised for sale in a London market as African in design; he later discovered the cloth was actually designed in the Netherlands, copying Indonesian batik. Although one might say that this bricolage of elements testifies to the global nature of our lives and contemporary creativity, Pennycook points out that we must not overlook how all these elements brought together make fresh meaning, a local assemblage that brings old elements from other places here, to form something new. This art piece, an emblem of creativity and iteration (or repetition), provides visual evidence of the powerful force of remaking (so that the same is also different). Pennycook's point is that if we look at any scene of local discourse, it is just such a rich palimpsest-in-process of old elements recombining to make something new. If we listen to language, he says, there are elements of other places, but they are being remade and localized here.

This view of language builds on the recognition that language is better understood as practice than as system. Systemic views of language have served regulatory purposes, introduced in order to convince users to follow an official procedure rather than allowing for the potential confusion of variety or the singularity and individualism of language experimentation. Scollon and Scollon remind us in a brief history that the *lingua franca* has not always been English, and that languages were not always regulated by rules and grammar. During the reign of Queen Isabella in Spain, they tell us, Elio Antonio de Nebrija prepared *Gramatica Castellana*, a grammar of Castilian Spanish to improve speech and regularize the conventions of writing, producing "the first grammar in any modern European tongue" (viii). Ivan Illich has referred to this as "a tool for conquest abroad and a weapon to suppress untutored speech at home . . . a system of scientific control of diversity within the entire kingdom" (35–36). The flaneur should pay attention to details of the actual linguistic landscape, perhaps even looking and listening for interruptions or transgressions that break the regulatory code. If language is a live practice, the flaneur is in a front line position not only to witness its local iterations, but to engage with others in active language production.

The multimodal flaneur not only looks at people, places, and print, but should also listen to the sounds of speech and noise to begin developing a sense of the city's rhythms. The concept of rhythm analysis was one that

Lefebvre developed late in his career, as a to way draw attention to pattern in the crowd, so that "one discerns flows in the apparent disorder and an order which is signaled by rhythms" (230). One street is not like another, and even taking a few steps can change one's sense of the mood of place. The flaneur can attend to these changing moods and atmospheres.

Walking somewhere new and paying attention to people, place, and signs, a practicing flaneur is not only prone to read a sign, but to assess whether people are heeding it, and even to consider who placed it and to what purpose. Rather than walking blindly through a daily routine, intent on getting from A to B in the most expedient way, the flaneur's strolling is without destination. By simply taking in the passing environment the flaneur begins to develop a critical consciousness, if we accept that non-perception and dependence on routine is for many the normative way of maneuvering through city congestion and maze.

MOBILITY: GLOBAL AND LOCAL

At a recent lecture about patterns of home and "unhoming" in children's literature, the speaker, a children's literature specialist, described the increasing reliance on the narrative pattern of hero homelessness. She pointed out that whereas the central character used to end up reunited with home, in contemporary stories, this figure often winds up homeless and often lives with nonfamily others, often in an unofficial support group that can be understood as an "assemblage" in place of traditional family (Reimer). As to why this narrative pattern may have changed, the speaker pointed to the influence of globalization in our thinking, with mobility as its key term. Money and people flow around the globe, and she speculated that children's narratives play a role in preparing young people to encounter the migratory challenges of contemporary culture.

If it is true that mobility is a key concept in contemporary life, then on this basis there is support for introducing our young adult students to flaneury, for it is a form of inquiry particularly responsive to mobility—a kind of knowing-in-motion. Yet rather than provide a premade narrative with actions and values that normalize particular forms of mobility in contemporary culture, flaneury requires the flaneur to actively engage in producing the map and network of movement and meaning. Agency is afoot! Moreover, flaneury recasts "home" in the public rather than domestic sphere, to the degree that it emphasizes exploring as one's own the civic community and urban environment. Thus the home space of belonging is made wider, as is the group to which one belongs, encouraging an expansive reframing of home as local environment and community.

Within flaneury, "movement" itself is a nuanced term. Rather than conceptualizing movement as something that takes one in and out of places—movement as facilitating simple entrances and exits—for the flaneur movements are multiple and simultaneous. The flaneur is part of the networks of motion, yet the self-consciousness required by the practice of this inquiry makes one aware of being apart from them. The standpoint is not stationary but always moving, so always verging on entry, but never fully committed to place. In this respect, the flaneur might be said to inhabit spaces on a temporary basis, as a liminal or marginal presence. Perhaps the best way to document how the postmodern flaneur can experience mobility and transient identity is to trace the history of flaneury from the late nineteenth century till now, noting that in place of a settled perspective, the postmodern flaneur has a transient identity.

If mobility, migration, and global flows are the standard fare of contemporary life, it is useful to enrich these terms by contextualizing them in the local environment. Practicing flaneury helps to develop an understanding of home as unexpectedly capacious and motion-filled, and of "mobility" as more than something that "just happens to people." The flaneur uses mobility for more than getting from A to B, not destination driven but on a learning journey. From this perspective we might say that the practice of flaneury rewrites the narrative of staying or leaving home by foregrounding a different set of questions about the networks of movement within place.

WORKS CITED

Araujo, Yara Rondon Guasque. "The City as Medium in McLuhan and Flusser." Trans. Carolina Siquera Muniz Ventura. *Flusser Studies 06*, May 2008. Web. 24 Feb. 2012.

Baudelaire, Charles. *Paris Spleen* [Prose poems]. Trans. Louise Varese. New Directions Paperbook: New York, 1970. Print.

———. *Tableaux Parisiens*. Trans. Print.

Birkerts, Sven. "Walter Benjamin, Flaneur: A Flanerie." *The Iowa Review* 13.3–4 (Spring 1983): 164–179. Print.

Calvino, Italo. *Invisible Cities*. Trans. William Weaver. London: Pan, 1979. Print.

Donald, James. *Imagining the Modern City*. Minneapolis: U of Minnesota P, 1999. Print.

Duhigg, Charles. *The Power of Habit: Why We Do What We Do in Life and Business*. New York: Random House, 2012. Print.

Dyer, Geoff. "Inhabiting." *Restlesss Cities*. Ed. Matthew Beaumont and Gregory Dart. London: Verso, 2010. 157–172. Print.

Kingwell, Mark. *Concrete Reveries: Consciousness and the City*. Toronto: Viking, 2008. Print.

Massey, Doreen. *For Space*. London: Sage, 2005. Print.

McLuhan, Marshall. *Culture Is Our Business*. New York: McGraw-Hill, 1970. Print.

McLuhan, Marshall, Kathryn Hutchon, and Eric McLuhan. *City as Classroom: Understanding Language and Media*. Agincourt, Ontario: The Book Society of Canada, 1977. Print.

Pennycook, Alastair. *Language as Local Practice*. New York: Routledge, 2010. Print.

———. "Linguistic Landscape and the Transgressive Semiotics of Graffiti." *Linguistic Landscape: Expanding the Scenery*. Ed. Elana Shohmay and Durk Gorter. New York: Taylor and Francis, 2009. 302–312. Print.

Reimer, Mavis. "No Place Like Home: Some Thoughts about Unhoming in Contemporary Culture." Presentation. U of Winnipeg, March 2012.

Reynolds, Nedra. *Geographies of Writing: Inhabiting Places and Encountering Difference*. Carbondale: Southern Illinois UP, 2004. Print.

Schaafsma, David, and Ruth Vinz. *Narrative Inquiry: Approaches to Language and Literacy Research*. New York: Teachers College P, 2011. Print.

Scollon, Ron, and Suzie Wong Scollon. *Discourses in Place: Language in the Material World*. London: Routledge, 2003. Print.

Soja, Edward W. *Thirdspace: Journeys to Los Angeles and Other Real and Imagined Places*. Oxford: Blackwell, 1966. Print.

Turner, Mark O. "Zigzagging." *Restless Cities*. Ed. Matthew Beaumont and Gregory Dart. London: Verso, 2010. 299–315. Print.

Part IV

Places of Resistance and Acceptance

12 From Concept to Action
Do Environmental Regulations Promote Sustainability?

*Becca Cammack, Linn K. Bekins,
and Allison Krug*

The crews look at the job site and they now notice when the (materials) don't seem to be installed correctly, and . . . there was some surprise that after a season of heavy rains, properly installed (materials) DO WORK. (case study participant)

INTRODUCTION

Over the last several decades, environmental regulations in the United States have grown in scope, complexity, and jurisdictional breadth. Although intended to limit the population's impact on our water, air, and natural resources, these regulations have also proven to be costly and restrictive in terms of commercial operations and development. Few industries today have been left untouched by the cost and effort involved in maintaining environmental compliance, with regulatory permits now widely issued throughout both the industrial and commercial sectors. As the means by which compliance is enforced and measured, environmental regulations are also the lever with which sustainability is transformed from idea into action, from concept into reality. Discussing the ecology of place in environmentalism in general, we ask, is *effective* environmental policy persuasive enough to make *place* irrelevant? In other words, can it subsume the particulars of place and become a matter of principle irrespective of place?

Persuasiveness hinges on concepts first promoted by Aristotle—the triad of logos, pathos, and ethos. In environmental discourse, Herndl and Brown propose that persuasion is effected by employing one or more of these qualities. In their model, science is anthropocentric (logos) presenting nature to be an object, its value quantifiable; the poetic, almost spiritual relationship we have with nature is eco-centric (pathos); and the role of the regulator to manage nature as a resource generally falls into the ethnocentric category (ethos). This model may help the academic sift through the mountain of literature on the topic—both scientific inquiry and in the popular press—but how does the general public respond to these various tactics? Are they equally persuasive? In this chapter, we examine the utility of ethos as an appeal to developers to embrace rules, to move the notion of sustainability from paper to practice.

In this review, the framework for place is the perspective of a small cast of environmental professionals in Southern California—managers at a public utility who are tasked with the responsibility of ensuring that crucial development projects comply with a specific environmental permit—one intended to prevent construction-related pollutants from entering storm water runoff. Two methods of analysis provide a comprehensive review of the dramatic tension between regulations and those who must comply: 1) a focused case study of those involved in conforming to the requirements traces the evolution of idea into action, and 2) a rhetorical analysis of the regulatory documents evaluates how language persuades us to accept rules and their rationale. Do environmental regulations actually help teach and encourage sustainable behaviors? To what degree are personal beliefs and lifestyle decisions influenced by the rhetoric of sustainability as it is communicated through environmental regulations? Our findings suggest that the regulatory process does indeed have the potential to promote sustainability, both on the job and off. Yet our industry insiders point to specific limitations that hinder regulatory efficacy—the transformation of sustainability from permit-chasing to the principles informing our daily choices.

ENVIRONMENTALISM COMES OF AGE

The postmodern environmental movement is entering its fifth decade of activism in a continuing struggle to create a culture of sustainability.[1] First emerging in the late 1960s in response to the rapid degradation seen during the Industrial Revolution (Rugman), the early environmental movement was based largely on protests and the formation of radical political groups. This movement, although young and sometimes volatile, represents the earliest organized attempt to combat America's seemingly careless management of pollutant sources, pesticides, toxic waste, and fuel-burning engines. Now, decades later, the movement is still waging war against these interminable foes. Only now, the battle tactics have changed. The grassroots efforts of the movement's infancy have given way to a more mature network of legislation and policy, with regulations now mandating the ways in which we use our water, air, and natural resources. Administered by a complicated network of regulatory agencies, environmental regulations mark the maturation and legitimization of a movement still seen as radical by some. And as we enter into a new decade where environmental degradation remains one of our most pressing and controversial topics, regulation can only be expected to grow more complex.

With the maturation of environmentalism, an entire discourse has evolved devoted to describing the who, how, why, what, and when of this movement. But just as the etiology of various psychological conditions entered the mainstream vernacular as "psychobabble," Goggin has noted that environmentalism has generated its own "sustainababble." The telegraphic language used in discussing environmental issues (Killingsworth and Palmer) creates a useful framework for discussion, but the rhetorical shorthand may obscure the potential for true collaboration by unnecessarily placing the parties in opposition, even if their philosophical views are actually rather closely aligned. We found this to be true in our observations of the regulatory agency and the public utility presented in this chapter—both have stakes in the ability of the utility's project to move forward and be successful. After all, the resulting development will ultimately deliver an important product to the public utility customer. Yet the process of securing a permit and adhering to its rules sets the stage for a polar dynamic often placing these two parties in opposition, caught in one of the "paralyzing dichotomies of ecospeak" (Killingsworth and Palmer).

In contrast, Killingsworth and Palmer's continuum of perspectives on the environment (11) places government and business as neighbors along the continuum, yet recognizes the tension inherent in this hegemonic relationship, one in which both parties are powerful. Thus, careful analysis of rhetorical appeals is crucial to casting regulations in such a way that they are more likely to evoke solidarity among the target audience. The tools of policy have the potential to create allies, and we see some evidence of this in our analysis. But much remains to be done, and this is also evident in our work. Those seeking to effect change run the risk of marginalizing their very existence unless they keep digging—not at the earth, but at the rhetoric that threatens to continue casting environmentalism as a fringe movement. This is Goggin's continued contribution to the rhetoric of sustainability—systematically soliciting evidence of its praxis. In so doing, he reminds us that this is a serious endeavor despite the many buzzwords and jargon threatening to co-op the notion of sustainability.

We contribute to this effort by applying the rhetorical scholar's unique perspective on the tools of policy to describe their transformation from bureaucratic rules into truly sustainable imperatives once they become part and parcel of everyday actions. In so doing, we present the methods of rhetorical analysis as a heuristic—to students of rhetorical analysis we introduce the instruments used to dissect and describe the documents employed in environmental law, and the degree to which these documents influence behavior. Although these methods are instructive, the narratives captured in our survey are given first billing. They will conjure an image of workers committed to adhering to rules; and despite the broad continuum of subscription to the rules, each participant demonstrates some understanding of their importance and relevance. The next time you drive by a construction site, we hope these specific examples of real people following the mandates of a simple permit enacted to manage storm water runoff will serve as a reminder that each of us should be so diligent wherever our paths lead. This would be evidence indeed of successful rhetorical appeal.

ENVIRONMENTAL POLICY—AN EVOLVING DRAMA

Environmental regulations—the tools used to maneuver the otherwise unwieldy concept of sustainability from idea into action—are enacted primarily in the form of permits issued by regulatory agencies (Altman). These permits are issued to both public and commercial entities for a variety of activities ranging from short-term development to ongoing daily operations. Over the last several decades, these permits and their regulatory origins have played an increasingly prevalent role in most commercial operations, with few industries left untouched by the cost and effort involved in maintaining environmental compliance.

Although enacted with the goal of restricting how individuals use and impact the earth's resources, many environmental regulations actually have very little jurisdiction at the residential level. Despite the impact that a single individual or family can have when choosing to live a sustainable lifestyle (or conversely, when defaulting to a wasteful one), many regulations and their related permits do not target the residential sector. Instead, the majority of regulations are written for and enacted within a target group considered to be more manageable from a regulatory perspective (and far more lucrative from a financial one): the commercial industry (Cropper). The commercial sector, the driver behind large-scale development and industrial operations, is also the most visible perpetrator of degradation—the pipes discharging oily wastes into rivers, stacks spewing clouds of gas into the atmosphere and generating hazardous wastes that are later disposed of in some less-than-sustainable way. In short, the commercial sector represents a highly visible source of environmental degradation. As such, this sector is also the target of hundreds of environmental regulations that have been enacted over the last fifty years.

The effects, both positive and negative, of this relatively young branch of regulation over the U.S.'s commercial industry has been a subject of debate since the 1980s when big business and industry first began to feel the sting of regulatory compliance. These debates were highlighted for the public in several journal articles, such as "The Challenge of Going Green," published in the *Harvard Business Review* in 1994. This article chronicled the debate among scholars and experts regarding the impacts of environmental regulations on American businesses. Although some experts have remained optimistic that environmental regulations have the potential to actually bring about cost savings as wastes are reduced and operations streamlined, many others have argued that the cost of doing business buried under a barrage of environmental permits presents a dangerous strain to American business (Clarke et al.). In either case, the proliferation of environmental regulations has only continued to intensify, and today nearly every segment of American business has been forced to adapt to the rising cost of compliance.

THE ENSEMBLE CAST AND SETTING

The Construction General Storm Water Permit (CGP) issued by California's State Water Resources Control Board (SWRCB) Division of Water Quality (Order 2009–0009-DWQ), is a regulatory permit issued to construction projects having > 1.0 acre of soil disturbance in California. Approximately 60,000 projects have obtained coverage under the CGP statute. The CGP is designed to minimize the loose sediment exposed and pollutants produced on job sites because the pollutants migrate from construction sites during rain and eventually reach storm drains and waterways. The CGP language is drawn from the Environmental Protection Agency's (EPA) own construction general permit for storm water, which applies similar permit coverage to >1.0 acre construction sites in states lacking a local branch of government for enforcement. The CGP, the EPA's general permit, and associated storm water regulations make the claim that sediment is one of the most prominent pollutants now found in bodies of water containing runoff and are primarily shed from construction sites and other non-stabilized surfaces located upstream.

Local infrastructure problems have also supported the evolution of storm water regulations. In San Diego County, specifically, the network of storm drains that drain the city, directing runoff to the ocean, no longer has the capacity to manage local runoff. Studies have shown that this drainage issue can be at least partially attributed to the accumulation of sediment and other debris in the storm drain system (City of San Diego), with the source of this sediment again tied to the construction projects that expose and dislodge soil. As a result, storm drains are backed up and unable to drain properly, leading to localized flooding frequently in low-lying and coastal areas, intensifying storm-related damages and adding strain to the already stretched municipal resources struggling to keep roads and properties intact.

The CGP provides requirements and places restrictions on how construction activities are conducted with the goal of minimizing the discharge of sediment and other pollutants from project work areas. When a construction project triggers coverage under the CGP, permitees submit notifications and fees to the SWRCB. The CGP also requires a permitted project to document site conditions and demonstrate that the work activities meet permit requirements. Permitted projects are required to develop project-specific plans describing the following activities: how water pollution will be prevented; onsite monitoring of construction activities; documentation that the storm water flowing off of the construction site is pollutant free; and stabilization of all disturbed areas at the end of the project (SWRCB). As employees of a public utility company in California, our case study participants were involved in all aspects of ensuring compliance with these requirements.

Conformance with these permit requirements often comes at a high price: cumulative costs for obtaining permit coverage, developing the required plans, and implementing each of the CGP's requirements throughout the lifetime of the project can add thousands of dollars to project costs. Many construction companies and land developers have struggled to remain in compliance with the seemingly exhaustive terms of the CGP, especially as it has grown more stringent with each renewal. Additionally, California's construction community has questioned the benefit of the CGP in actually protecting water quality. Critics of the permit have questioned how or if it does anything at all to help or protect the environment, especially in the face of government budget cuts which have limited the resources available for consistent enforcement. Our review of the language used to justify the CGP and the perceptions of our cast of environmental professionals shed light on this debate.

CREATING A CULTURE OF SUSTAINABILITY

This chapter trains the rhetorical lens on the concept of sustainability as it is communicated through environmental regulations—in this case a specific regulation involving a permitting process ultimately intended to prevent the release of pollutants into storm water runoff. We focus this essay first on the public utility interviews (the case study) because their response to the document illustrates the degree to which the policy tool is an effective rhetorical appeal. Thus, to test conveyance of concept to action, a small case study was conducted involving a group of environmental professionals who together have more than fifty years of active experience working within the environmental industry, and within the restrictions of the CGP and other regulations. The goal was to determine the extent to which these individuals had incorporated sustainability into their consciousness, decisions, and daily behaviors as a result of working within the constraints of environmental regulations. Had their values or views on sustainability changed as a result of their exposure to the permit? Have they reconsidered how their work or personal actions can impact the environment? Or does the CGP simply prescribe a set of rules to people who have no choice but to conform?

The central question here is whether or not environmental regulations enhance the responsible management of resources. Do environmental regulations actually promote and encourage sustainability across the continuum of place—job site, residence, community activities? To this end, we specifically examined the most proximal impact of state regulations—the enhanced awareness and deliberate actions taken by those responsible for conforming to the storm water permit. While the surveys were in the hands of our participants, we turned our attention to the policy tools. Utilizing a systematic rhetorical analysis of the CGP permit and accompanying Fact Sheet, we evaluated the methods used to engender allegiance to the concept of "sustainability." We present our methods and findings from both analytical endeavors here, starting with the case study.

CASE STUDY: THE POTENTIAL FOR BROAD INFLUENCE

> These are the most likely people to support environmental initiatives. I wonder if they don't care or don't understand. If they are not going to follow the rules is there any hope? Not sure if sustainability is possible but I do think we (humans) need to try to work toward sustainability. (case study participant)

To gauge the CGP's role and degree of influence in the lifestyles, behaviors, and attitudes of people who work closely with it, we spoke with four utility environmental professionals. Each has worked extensively on construction projects covered under the CGP, and their related experiences attest to the influence of the permit in real life. With a combined experience of fifty-four years, these professionals were selected for the study based on having high-level experience working with environmental regulations and construction projects, and their specific exposure to construction projects involving long-term management of the CGP.

Thirteen survey questions pertaining to the CGP and its impact on attitudes and behaviors were developed, tested, and distributed to the case study participants. The questions prompted participants to provide in-depth responses surrounding the influence of the permit on mindsets and behaviors, in addition to the mindsets and behaviors of other people they have observed. To assure unbiased and complete reporting, the participants were offered the opportunity to respond anonymously if they wished. We were interested in whether the people working within the framework of these regulations begin to view the world and how they impact it differently. Do they incorporate new, more sustainable actions into their lives as a result of being thrust into the mindset of regulatory compliance? Do they avoid choices and actions with the potential to harm the environment because of their exposure to these regulations? Or, conversely, do environmental regulations simply prescribe restrictions and requirements that mandate how work is carried out, without having any significant educational or persuasive sway?

Generally, participants either incorporated a new practice into their lives (or eliminated an old one) as a result of their exposure to the CGP. The participants who disagreed also stated that as long-time members of the professional environmental community, they had already incorporated sustainable behaviors into their lives. Additionally, over time the participants witnessed a change in perception, both in themselves and others, toward the CGP. Generally, a gradual acceptance and sense of understanding seemed to take place.

A culture of environmental awareness is clearly evident among those surveyed. One participant observed that she now sees issues with water quality everywhere she goes. She talked of the evidence she observed of damage to trails and plants while out hiking in areas where, presumably, the other hikers would be nature lovers and somewhat concerned with sustainability and environmental quality. However, the degradation that she has witnessed to the resources in these areas has discouraged her. Others indicated that although they were already active members of the environmental community and aware of environmental issues, the CGP has advanced a more thorough awareness of issues related to storm water runoff and pollutants. Only one of the participants declared that the CGP had not substantively changed her habits; however, she acknowledged that as an environmental professional with many years of experience in her field, she already felt that she was aware of related issues. Citing specific examples of changes in personal behaviors, one of the participants listed the following conscious decisions: the use of protective measures in her own yard to retain soil and prevent it from being discharged to storm drains; avoidance of washing the car in the street; staying on the trail while hiking; and avoidance of off-road driving. Other participants indicated that they were already working on reducing their ecological footprints and the CGP had not directly influenced any lifestyle considerations.

Although supportive of the CGP principles, these participants astutely note that many of the permit requirements prescribed in the CGP are perhaps too stringent and, ultimately, are extraneous when it comes to protecting water quality. This perhaps reflects a disconnect between the document and its delivery. Ultimately, the CGP would be better served by taking a multidisciplinary approach that folds in other types of environmental issues. One participant stated that "as a biologist it really bothers me that the environmental agencies are so siloed. I would like to see some biology requirements in the re-vegetation requirements in the permit." Another participant was initially skeptical of many of the CGP's requirements, but now acknowledges that there is value in providing clearer guidelines and rules for workers to follow. Heartening, perhaps, is the possibility that the administrative burdens of managing the permit throughout a project's life cycle are offset by improved awareness among all job site workers:

> The crews look at the job site and they now notice when the (materials) don't seem to be installed correctly, and for some road grading work there was some surprise that after a season of heavy rains, properly installed (materials) DO WORK.

Nevertheless, the response to the CGP on the jobsite has been mixed. Although the participants indicate that construction managers generally accept the CGP, they are uncertain of the degree to which this acceptance is motivated by expediency (just get the job done) or true environmental concern. Explaining the permit process and what the CGP is trying to accomplish does seem to soften the blow:

> Some of the construction personnel I work with have had slightly "frustrated" attitudes with some of the additional obligations we have

as a company. Explaining the permit helps them understand where the requirements come from and help soften their attitudes.

In the end, does the CGP promote sustainability? Among the managers we surveyed, the consensus was, at best, doubt with respect to its utility in this regard. Participants generally felt that the CGP is too specialized; with a specific focus on construction-related pollutants and water quality, it leaves other environmental issues unaddressed and does nothing to encourage comprehensive environmental protection. One participant specifically noted that the CGP does nothing to actually preserve natural resources because "in no real way does it slow/limit development." The day-to-day job demands may obscure our general understanding of how the requirements fit into the larger concept of sustainability. Making this leap is not possible for many people: "For many, the leap from reducing, managing, controlling storm water runoff to protect the downstream environment to assisting in the sustainability of the local environment is too large of a leap to make." Additionally, participants felt that the CGP may only teach sustainability to those working at higher project or programmatic levels, or those who have already been somewhat educated in matters of the environment and sustainability. From this perspective, people outside of the environmental bubble (to include laborers and crew members actually involved in construction activities and soil disturbance) may not perceive the same messages in their own exposure to the CGP.

Those we interviewed clearly thought the potential for education and promotion of sustainability is there, however generally agreed that the focus of the CGP is too narrow. Better integration with other agencies would not only ensure the prevention of pollutants from entering runoff but also stimulate sustainability by guiding other development activities, such as long-term runoff control through biological means and sensitivity to culturally meaningful areas. To embrace the notion of sustainability in a meaningful (and sustainable?) way, permit users would have to educate themselves by some other means outside of the permit. In a sense, the permitees must still have an internal motivation to see the larger context. The agency should have "greater focus on looking at where storm water runoff might have an actual impact on the environment, and not be as concerned with making sure that everything is in place in those areas less likely to be impacted."

In summary, these responses suggest that the CGP does, in fact, promote sustainability among a subset of the population already so inclined. But the agency needs to address the disconnect between the document (the permit and its rules) and its delivery (how the message is translated on the job site). The systems thinkers "get it" whereas those focused on getting the job done may need some help making the conceptual leap. Perhaps a more compelling case would be made by surveying those closer to the shovels—the hardworking construction personnel subcontracted to do a difficult 10-hour-a-day job. Additional research across a broader demographic would clarify the degree to which place influences the assimilation of environmental regulations.

RHETORICAL ANALYSIS

In parallel with the qualitative component of this research project, we conducted a rhetorical analysis to identify the presence and use of persuasive elements within the content of the CGP. This analysis shows how persuasive language in the CGP encourages the target audience to accept a sense of environmental responsibility, and ultimately to adopt more sustainable lifestyles. Collectively, these two efforts allowed us to identify the use of rhetoric in the CGP, and to describe the degree to which the CGP holds rhetorical power in practice.

The CGP Fact Sheet was prepared by the SWRCB to explain and justify the numerous revisions and new requirements proposed for the existing version of the CGP, which was renewed in 2010. These revisions and requirements proved to be a controversial topic of discussion among members of California's construction industry in the months (and even years) leading up to the current CGP's adoption. Many in the community viewed them as extreme, costly, and at times even irrelevant in terms of protecting water quality. It was also widely acknowledged that the revisions proposed for the CGP would continue to escalate the costs of development in California, and that compliance with the related permit requirements would require more time, energy, and accountability on the part of the builders.

The Fact Sheet was prepared by the SWRCB as an effort to address the comments and concerns raised by the construction community to the proposed changes. Contrary to what its name implies, this lengthy forty-nine-page document is a clear demonstration of persuasive language, from the formal document construction (comprised of a "Background" and "Rationale") to the provision of nine tables and six figures filled with scientific data. We begin our summary of rhetorical tactics in the Background section, which reviews three decades of regulatory evolution in a clear attempt to validate the legitimacy of the CGP, with emphasis on the Clean Water Act (CWA) and its regulatory lineage over the last thirty years. Subsections include a review of legal challenges that ultimately recognized storm water as a point source for pollution and brought it into focus as a central component of water quality law. A "Blue Ribbon Panel" of experts was convened to discuss the feasibility of actually measuring the pollutant levels in effluent leaving jobsites (one of the measures initially proposed and then later adopted into the CGP) and providing data as part of the permit process. The CGP thereby suggests to the reader that the panel is comprised of experts in the field who should be accepted as the "best of show" and thus credible. The direct quotations from the panel lent validity to their findings, as the findings appear to be official. The quotes also project ownership of findings onto the panel versus the SWRCB, which potentially diffuses criticism.

The second section of the Fact Sheet, the "Rationale," summarizes the adopted changes in a concise list, item by item. Significant changes are explained in detail, particularly those pertaining to the proposed numerical effluent limits or other major impacts to project design and execution practices. Accompanying figures and tables are used to illustrate the scientific data backing the changes, further justifying them. The "no-nonsense" presentation of the new requirements and incorporation of data—such as graphs and figures which serve to visually depict the processes where raw data could not—further supplement the rhetorical framework of the Background section's lengthy narrative. The use of data serves to "seal the deal" rhetorically, silencing all objections with the simple presentation of logic.

Unlike the fact sheet, the CGP itself does not incorporate attempts to explain or justify its restrictions and requirements. This is to be expected, as the CGP serves a completely different function than the Fact Sheet. As regulation, the CGP prescribes rules, restrictions, and requirements that are presented systematically, section by section. Similar to the Fact Sheet, however, the CGP opens its section on General Findings by discussing the Clean Water Act (CWA) and how storm water became a part of this federal water quality law. This opening reference to the CWA is an astute rhetorical move as it bases the contents of the permit on the well-established EPA and CWA. These two references give the CGP the appearance of being more robust, grounded within a network of federal law and regulation.

Even with its grounding in technical and regulatory terminology, however, the CGP does make use of footnotes to provide similar justifications and attempts to explain revisions and new requirements (similar to the Fact Sheet). These footnotes provide a means for explaining the basis of the new requirements introduced in the current version of the CGP. Other rhetorical tools used in the CGP involve the use of strong action words such as "require," "limit," and "prohibit." By clearly identifying permit requirements, limitations, and prohibitions, permitees are clearly informed of what they can and cannot do. In addition, the CGP language represents that of a standard operating procedure in that it clearly lays out directives by informing permitees how they "shall" fulfill permit requirements.

In summary, our rhetorical analysis identified the presence and use of rhetoric primarily in the Fact Sheet, a document that was prepared to accompany and justify the CGP. In contrast to the CGP itself, the Fact Sheet provides an in-depth background discussion on storm water laws and relevance to more general water quality issues. This discussion starts with the original implementation of the CWA, and goes on to narrate its evolution as the significance of storm water pollution was gradually recognized and then incorporated into water quality law. This association with the CWA and federally recognized court decisions is effective rhetorically as it provides an official grounding for the CGP. And, conceptually, the CGP transitions from a set of frustrating rules administered by local enforcement agencies into an important contributor toward a global initiative: the protection of our water quality.

The Fact Sheet also makes use of supporting scientific data in the form of graphs and tables, further validating the new requirements and restrictions incorporated into the 2010 revision of the CGP. With California's construction community viewing many of the proposed changes as extreme, costly, and even somewhat irrelevant to the overall protection of water quality, the Fact Sheet's use and presentation of scientific data seek to link the CGP to larger environmental issues. When presented with visual representations of processes such as erosion and sediment loss, the proposed changes are perceived more as a necessary means to an end (as opposed to a senseless and costly network of rules). In short, the Fact Sheet uses scientific data, a connection between the CGP and long-established federal water quality laws, and the logical connection between strict work practices and environmental conservation to persuade its users to accept the CGP.

CAREFUL CONVEYANCE OF THE CONCEPT: CAN IT CREATE A CULTURE OF SUSTAINABILITY?

Just as properties are conveyed from prior owner to new owner, so is the concept of sustainability as it ages. Those involved in executing the principles of sustainability are concept owners, and careful conveyance of the *concept* of sustainability is essential to the creation of a citizenry that embraces the *culture* of sustainability. Only when the concept becomes culture is it truly sustainable. As the environmental professionals point out, making the leap from permit to practice is highly variable. To some extent, probably, the CGP actually does have rhetorical power—it is persuasive in promoting sustainability at least in the context of the job site. And certainly on paper, the CGP makes use of a wide variety of rhetorical tools, relying strongly on the visual presentation of scientific data and a connection with over thirty years of federal water quality laws to justify its constraints. In practice, it appears that sustainable choices are made both on and off the job site, and many destructive behaviors are eliminated in light of exposure to environmental regulations.

Even when taking criticisms into account, it can be concluded that (both on paper and in real life), environmental regulations have the potential to promote sustainability. And if these regulations are imparted properly, sustainability can be transformed from a concept into deliberate actions, behaviors, and choices regardless of the particulars of place. It can become part of the culture of a particular place. The job site, its workers, and its neighbors are the epicenter of sustainability demonstrated. The next step should be gathering specific input on what the "proper" conveyance of sustainability principles might look like. We must first ascertain which rhetorical tools are most persuasive according to audience characteristics (managers, workers, consumers), then optimize our delivery by testing a variety of media to carry the messages. Certainly there is room for additional research into this question to inform policy makers, code writers, developers, and citizens.

NOTES

1. Elusive "sustainability." The concept of sustainability has evolved in the last two decades, yet remains somewhat elusive because it means a variety of things to an even wider variety of users. Is it a clear scientific goal, a call to action, or a journey? Two definitions have informed the postmodern environmental movement:

   a. Those actions that "meet present needs without compromising the ability of future generations to meet their needs." (UN Documents)

   b. Sustainability is based on a simple principle: Everything that we need for our survival and well-being depends, either directly or indirectly, on our natural environment. Sustainability creates and maintains the conditions under which humans and nature can exist in productive harmony, that permit fulfilling the social, economic and other requirements of present and future generations. Sustainability is important to making sure that we have, and will continue to have, the water, materials, and resources to protect human health and our environment. (EPA)

WORKS CITED

Altman, Morris. "When Green Isn't Mean: Economic Theory and the Heuristics of the Impact of Environmental Regulations on Competitiveness and Opportunity Costs." *Ecological Economics* 36 (2001): 31–44. Print.

City of San Diego (California). *Department Description: Storm Water.* San Diego, CA, 2008. Web. 26 Jan. 2012.

Clarke, Richard A., Robert N. Stavins, J. Ladd Greeno, Joan L. Bavaria, Frances Cairncross, Daniel C. Esty, Bruce Smart, Johan Piet, Richard P. Wells, Rob Gray, Kurt Fischer, and Johan Schot. "The Challenge of Going Green." *Harvard Business Review* 72.4 (1994): 37–48. Print.

Cropper, Maureen L., and Wallace E. Oates. "Environmental Economics: A Survey." *Journal of Economic Literature* 30.2 (1992): 675–740. Print.

EPA (United States Environmental Protection Agency). *Sustainability,* n.d. Web. 26 Jan. 2012.

Goggin, Peter, ed. *Rhetorics, Literacies and Narratives of Sustainability.* New York: Routledge, 2009. Print.

Herndl, Carl G., and Stuart C. Brown, eds. *Green Culture: Environmental Rhetoric in Contemporary America.* Madison: U of Wisconsin P, 1996. Print.

Killingsworth, M. Jimmie, and Jacqueline S. Palmer. *Ecospeak: Rhetoric and Environmental Politics in America.* Carbondale: Southern Illinois UP, 1992. Print.

Rugman, Alan M. "Corporate Strategies and Environmental Regulations—an Organizing Framework." *International Business: Critical Perspectives on Business and Management.* Ed. Alan Rugman. London: Routledge, 2002. 301–324. Print.

State of California. State Water Resources Control Board. *Construction General Storm Water Permit.* Sacramento, CA, 2010. Web. 26 Jan. 2012.

"UN Documents: Gathering a Body of Global Agreements." *Report of the World Commission on Environment and Development: Our Common Future,* n.d. Web. 26 Jan. 2012.

13 Mapping Literacies
Land-Use Planning and the Sponsorship of Place

Rebecca Powell

> Comment: I would like to publicly comment on the process of Vision 2040. I was immensely disappointed in the process guiding the Round 2 meetings for Vision 2040 that I attended earlier this year . . . In sum, I would like a process that is built around relationships, conversation, and in-depth choices. The last meeting felt like I was in fourth grade with my teacher telling me exactly what to do because I didn't know any better. (Appendix 2, Public Input, Citizen e-mail, 133)

The above citizen comment encapsulates the conflicts in the story of community planning: experts come to town, residents critique the experts, and the plan or process advocated by the experts never reaches fruition. It is a story repeated in communities across the nation. In a 2011 survey success of planning initiatives in Maryland, the National Center for Smart Growth Research found that although many planning initiatives have gone forward, few have reached fruition (Sartori et al. 2). The story's unsatisfactory conclusion (the abandonment of the planning process) is often attributed to a mistrust of outsiders and a conflict between expert and local knowledge. I want to turn attention to the rhetorical and place-based dimensions of this story as it unfolds in the land-use planning process of Doña Ana County, in order to suggest that the story's repetition is not due to a widespread distrust of outsiders, but a form of literacy and rhetorical critique based on local knowledge. This reframing of the narrative encourages academics invested in environmental advocacy and community literacy to consider lived experience and the complexities of location as resources for literacy, civic participation, and rhetorical interventions.

Environmental discourse and community literacy collide in the land-use planning process. From the gridded streets of a city's original plat to the city pool with the Works Progress Administration plaque, the history of land-use planning permeates the lives of ordinary citizens in the U.S., creating opportunities and obstacles. Historically, land-use planning has mostly happened sporadically on the whims of government and developers, exemplified by the large-scale efforts of the government to settle the West through the Homestead Act of 1862 and the smaller, but long-reaching effects of developers like Levitt and Sons' Levittowns. Currently, land-use planning practices have adopted the "smart growth" strategy that hopes to direct future growth to in-fill and protect environmentally critical areas (Knaap 4). In the last fifteen years, a growing numbers of communities have adopted smart growth strategies to plan future development, as exemplified in this chapter by regional planning efforts (Vision 2040) in Doña Ana County, New Mexico. To accomplish its goals of directing development toward existing communities and encouraging community and stakeholder collaboration, land-use planning asks for public input, putting community members, developers, and public officials in dialogue (Knaap 9). Thus, land-use planning is an overt example of critical geographer Edward Soja's claim that "[t]hroughout our lives, we are enmeshed in efforts to shape the spaces in which we live while at the same time these established and evolving spaces are shaping our lives in many different ways" (71). These spaces both shaped by, and shaping our lives, are the subject of the land-use planning process.

This chapter extends the growing scholarship on the uses of literacy in environmental arguments (Goggin and Long; Mason; Lindgren). By examining what resources participants deployed to influence the shaping of the environment during the land-use planning process and how the environment shaped the available resources, this chapter claims that lived experience and the complexity of location work as literacy resources and rhetorical opportunities for civic participation. Civic participation in the land-use planning process can be understood taking part in a literacy event as explained by Anne Whiston Spirn, a landscape architect and scholar, who equates effective land-use planning with reading the landscape, linking land-use planning and literacy (400). Unlike traditional models of literacy (reading and writing print-based materials), reading the landscape acknowledges lived experience and the complexity of location (Schroeder 279). Like literacy, for land-use planning to be effective, it must read and interpret local conditions, both of the environment and the people. These readings and interpretations are not made in isolation, but contested in public forums and scrutinized throughout the land-use planning process. The contestations and scrutiny take written and oral form, making the land-use planning process a rhetorical activity dependent on an awareness of location and environment with high stakes for all involved. These literacy resources and rhetorical interventions often take the form of critique and may shut down environmental discourse if the dialogue ignores conflict in favor of consensus, the end rhetorical goal.

My approach for this study includes textual analysis and participant observation. The primary data consist of the public comment section, Appendix 2, of the final draft of *One Vision, One Valley*, the name of the resulting document from the Vision 2040 process, and field notes and texts compiled from two land-use planning meetings in Las Cruces, New Mexico. Appendix 2 documents public input over the four-year process, specifically what the planners call "the public participation program" (23). It includes results from 1,600 surveys, public comments during workshops, meetings and comment periods, and 160 workbook tabulations from an early meeting. The second body of data, field notes and texts, were compiled at two public meetings I attended, an Advisory Committee meeting and a citizen's meeting. The goal of these meetings was to "present future development concepts for the City and County based on encouraging development within 'nodes' or neighborhood centers, as well as a land use model that will be used to compare alternative development scenarios" (Vision 2040 website).

The Doña Ana case illustrates how the complexities of location influence the planning process as participants draw upon local literacies and knowledge to take part in shaping their lived environment. To analyze how participants influenced the planning process, I ground the discussion of the process within the setting and exigency of Doña Ana County, a setting and exigency that holds certain literacies, visions of public participation and definitions of appropriate behavior. I explore how planners and presenters constrained participation and how participants drew upon locally gained knowledge to work through those constraints and influence the planning process.

THE SETTING, EXIGENCY, AND PLAYERS

Doña Ana County is the second largest and fastest-growing county in New Mexico with a population of a little over 225,000 people (Clapp 36). Thirty-eight hundred square miles of high desert on the border of Mexico and Texas, Doña Ana County's population has grown over eleven percent in the last six years, a rate almost twice that of the state's (U.S. Census). Bisected by the Rio Grande River, Doña Ana County is also home to a large agricultural sector. Studies predict continued population growth as industry, retirees and job seekers take advantage of the area's high desert climate, unique border location, military industrial complex, land grant university, and the installation of an international spaceport (Clapp 37). Located between the banks of the Rio Grande River and the Organ Mountains, the county seat, Las Cruces, was ranked in the top small metro areas for business, careers, and retirement by Forbes, the Milken Institute, CNN Money, and AARP over the last six years (MVEDA par. 1). This national attention has helped Las Cruces double its population since 1970 (par. 2).

In response to this population growth, Doña Ana County and the City of Las Cruces began a regional planning project (Vision 2040) in 2007, funded in part by a grant from the state (*Vision 2040*). A seven-step process, Vision 2040 is a collaborative city/county effort to create a comprehensive plan to serve as a legal basis for zoning and future land use within the city and county. Originally, planners thought the planning process would take three years; at this time, the plan is going into its fourth year and projected to be completed by the fifth or sixth year. Currently, six separate entities oversee development and zoning within Doña Ana County, operating under different and sometimes conflicting plans. In Las Cruces alone, the county, city, and extra-territorial zoning commission oversee zoning and building requirements. Vision 2040, in the words of a planning consultant, seeks to "make plans talk to one another" (field notes). This comprehensive plan will need to persuade diverse stakeholders (the six planning entities, developer and citizens).

To facilitate the process of Vision 2040, the city and county hired planning consultants, Peter J. Smith & Co. of Buffalo, New York. According to their website, Peter J. Smith & Co. specializes in urban design, urban and regional planning, and economic development. The consultants are responsible for all seven steps of the process, including developing and executing the public input process, creating existing and future land-use plans, and creating the final document (*Vision 2040*). Three consultants, one man and two women, attended the land-use meetings. (Hereafter I refer to them as the "presenters").

Other major players in the Vision 2040 process include county and city planning officials, the advisory committee, focus groups, and ordinary citizens. The county and city planning officials provide the consultants with technical support, access to the community, key contacts, and meeting venues. They also act as a liaison to the city council and county commissioner. The advisory committee members, made of hand-picked community and business leaders, play the complicated roles of cheerleaders, guinea pigs, and liaisons to the community. Members consist of a former mayor, a bank president, the president of the farm bureau, and other community and business leaders. Citizens not named to the advisory committee are also an integral part of the Vision 2040 process. Their expertise has been tapped for focus groups and data collection (*Vision 2040*).

When I attended two meetings, Vision 2040 had already completed the first two steps of the process, data collection and visioning sessions. The meetings I attended presented two land-use model maps, derived from the data collection and visioning sessions, projecting growth until 2040. The first map, "Business as Usual," shows what growth would look like if current planning practices continue. The second map, "Alternative Land Use," shows what growth would look like if the city and county adopted smart growth measures. As noted above, Appendix 2 of the final draft document gives a more complete picture of public participation, documenting public comments and participation throughout the entire three-year process.

CREATING CONSENSUS, IGNORING CONFLICTS

Before the public participation program could begin, specific publics needed to be created. Although rhetoric can create affiliations (Burke), some publics are created not through rhetoric, but by institutional structures. Publics created through institutional structures, or in this case planning mechanisms, have an added rhetorical burden when working in active society. Gerald Hauser defines an active society as one "in which citizens have differences over resources, contest their control, and include structures that encourage and have the capacity to achieve and perform negotiated resolutions" (254) and that "rhetoric serves as such a society's inventional resource for establishing relations by which it continually produces itself" (235). This definition of the public sphere emphasizes difference, conflict, and the relations that set up and negotiate differences and conflict. Importantly, Hauser sees rhetoric as the generator of those relations, relations that keep the society alive, for good or bad. Thus, publics created by planning mechanisms must forge their own relations through identifications and other means, while already being classified and acting as a group. During this land-use planning process, institutionally formed publics coalesced around status and common interests, but the underlying conflicts of these publics, conflicts based on their different experiences and visions of place, were not given space to be negotiated These conflicts did not disappear, but reappeared in the final draft comment stage, lengthening the land-use planning process by a year. The following examples illustrate how place-based conflicts were ignored and side-stepped by planners, who wanted to keep the process moving.

In the beginning, the planners established a three-tier system of participation: the advisory committee, focus group members, and public meeting participants. Both advisory committee and focus group members were invited participants. The Vision 2040 website does not explain how the advisory committee participants were chosen, only that they were invited and belong to various organizations within the community. However, focus group participants were invited to volunteer during the early stages of the planning process. Citizens who were not a part of the advisory committee or the focus group could attend public meetings, seven rounds of which have been held to date. Public meetings were held at varying times and locations. Moreover, 2,000 citizens had an opportunity to participate in a phone and mail survey about their land-use priorities and impressions of the region.

This three-tier system of participation that established conflicting relations based on how they viewed place, land, and resources both created and limited the types of publics in the planning process. For example, the seven-person steering committee of the advisory committee represented major economic interests in the community (agriculture, real estate, education, military, and banking), but it did not represent major societal and cultural interests in the community (social services, the arts, natural resource conservation, etc.). This division between place as economic resource and place as communal property haunted the planning proceedings. Moreover, a steering committee representing only economic interests, sometimes *opposed* economic interests, created relations that would need considerable rhetorical practice to negotiate.

Likewise, the focus groups were organized around set categories (agriculture and ranching; arts, culture, and history; citizens and neighborhood organizations; conservation and recreation; cyclists; housing and social services; real estate development; transportation; utilities and infrastructure). These focus groups were comprised of people with different lived experiences of place. For example, the agriculture and ranching focus group was composed of people who lived along the Rio Grande or the Sierra de Las Uvas, a mountain range on the northwest side of the valley, whereas conservation and recreation focus group members lived on the east side of the valley, underneath the shadow of the Organ Mountains and BLM land, used for the grazing of cattle and recreation. Although these locations may seem minutely different to an outsider, for the members of the focus groups, they created different priorities for the land-use planning process, priorities that would lead to long-term conflict. Focus groups were used to gain "specialized" knowledge of the area through guided conversations. The arbitrators of these conversations were the planners, and it was up to the planners to negotiate the competing specialized knowledges and locations of the area to make a rhetorically persuasive document. Conflicts that have existed between the groups and their specialized knowledges of the area were not given space to be negotiated by actual participants, participants who had a stake in producing the active society of the area. Instead, those conflicts were either ignored or synthesized by the planners in the creation of the document, a move that would lead to an extension of the process by a year and a voluminous public comment period where participants tried to re-articulate those conflicts in writing.

In both the advisory committee and the focus group sessions, participants articulated different visions and priorities of the area. For example, in the visioning meeting of the advisory committee meeting, when asked to describe the character of Las Cruces and Doña Ana County, five members replied "rural," whereas seven replied "growing and metropolitan." These differing answers were not discussed, but left to the planners to incorporate. In another instance, when asked about their vision for the region in twenty years, some saw a "well-organized, well-planned" region with economic opportunities, whereas others concentrated on the "preservation of agricultural resources" and "preserved natural resources," comments that the planners grouped in the same category, although in an arid region agriculture and natural resources are often vying for water and land. Again, the planners did not open these seemingly conflicting responses to discussion; instead they grouped responses together and tried to make a vision statement encompassing a rural, growing metropolis that wants to be well-organized and well-planned, have economic opportunities, while preserving agricultural and natural resources.

To avoid conflict and create buy-in to the process, the planners also limited access to information. At the advisory committee meeting, participants were given five packets of information: Method and Breakdown

Explanations, twenty-two pages with no page numbers; a one-sided Agenda Sheet containing nine items; a Community Survey Results packet of five pages with page numbers and titles; a two-sided colored Future Land Use Scenarios, explained as shortened goals derived from survey; and ten pages of colored maps without page numbers. In addition, the PowerPoint presentation worked as a text. The availability of texts positioned the advisory council participants as people who make correct interpretations. However, these "people who make correct interpretations" asked a lot of questions and debated many of the terms and specifics of the maps. In light of this, a very different scenario unfolded at the public meeting, a week later.

At the public meeting, citizens were handed a two-sided black and white Future Land Use Scenarios sheet. The five packets from the week before were missing. Citizen's access to data was restricted. I assumed the lack of texts at the citizen's meeting reflected the costs of copying colored documents, but when I asked the senior city planner why the texts were not available, he replied, "We didn't think they were helpful in getting our message across last night." Planners and presenters were worried about differing interpretations of the data that arose during the advisory committee meeting. They were worried about conflict. What was important to the planners was getting the message across and continuing the process of public participation.

The planners created groups with different experiences of the lived environment. Not only did they live in different areas of Doña Ana County, but they also had different economic and communal relationships with the area. They were destined to conflict, the mark of Hauser's active society. However, the conflict obscured the planners' goal of buy-in to the planning process, so they took steps to ignore and side-step areas of conflict. Although this achieved the short-term rhetorical goal of delivering the message, it belied the long-term goal of creating consensus around the final document, a goal that is still currently delayed.

DELIVERY: ASSIGNING AND ASSERTING

As participants brought their different lived experiences of the valley to the planning process, similarly the planners and presenters brought their expectations and previous experiences to the planning process. These expectations and previous experiences surfaced in the spatial configurations of the meetings, circulations of texts, and the chosen modes of delivery. Both meetings took place in a conference room of the Doña Ana County Courthouse. During the advisory council meeting the participants were seated at tables arranged in a horseshoe configuration. The presenters sat at the head of the horseshoe, beside a projector, in front of the drop-down screen. Easels holding colored maps lined both sides of the screen. The configuration of the room projected a workshop-type atmosphere with participants and presenters on equal footing. During the citizen's meeting in the same room, the tables were removed, and twelve rows of straight-backed chairs faced the screen and easels. The presenters stood against the wall during the presentation, waiting for their time to speak. The configuration of the room positioned the audience as receivers of knowledge and the presenters as givers of knowledge. In both meetings, the spatial configuration worked to reinforce the identity of participants and presenters. In the advisory committee meeting, the workshop configuration confirmed the advisory committee members' position as chosen team members. The lecture configuration of the citizen's meeting positioned participants as onlookers.

Both meetings began with a PowerPoint presentation. Although the presenter frequently asked for questions, neither the advisory committee members nor the citizen participants submitted any questions during the PowerPoint presentation. The PowerPoint positioned the presenters as performers of knowledge, asserting their expertise and message. Educator Catherine Adams notes that "PowerPoint reifies the notion of teaching as presentation, not conversation" (403). In this instance, PowerPoint allowed the presenters to display a persuasive case for the maps without having to discuss the ideas. While it was running, PowerPoint effectively quashed active discussion and debate, achieving the presenters' goal of broadcasting the message.

Yet what probably seemed like an expedient way to deliver a message to the presenters became an access point for participants. In written comments, several participants critiqued the delivery and used their critiques to leverage their own ethos as citizens. For Hauser, ethos hinges on performance, the performance of identity. Hence, Hauser defines the credibility of the citizen, "The ethos of the free citizen is made legible through exhibitions of self-worth" (234). A citizen is seen as a citizen, one worthy of participating in democratic processes, through a performance of "self-worth. One participant wrote,

> The reading of the entire PowerPoint aloud seems . . . redundant? The group completely lost interest. I believe we are all probably capable of skimming the packet and slides and understanding all that it contains. It's a waste of people's valuable time, time we're giving because we care. (*Vision 2040*)

The participant delivered her critique, but also managed to include an estimation of her self-worth (her valuable time and concern for her community). Later in this process, this same citizen went on to be part of a focus group. Although the environment of the meetings was controlled by planners, a rhetorical critique of the delivery and environment allowed participants to influence the process.

Historically, the ancient canon of delivery referred to the stance, gestures, and tone with which one delivered a speech, but recently delivery has experienced a revival. Prior et al. suggest that when we speak of delivery, we speak of distribution (how the message will circulate) and mode (what material/oral shape the message will take and in what context) (6). This expanded definition of delivery attends to the spatial configurations of meetings, circulation of texts, and the shape that the presenters chose to deliver their message. In this case, participants were assigned positions through the configuration of space and the availability and use of texts. Participants used critique of the delivery choices of the presenters to renegotiate those assigned positions and assert their authority over the process. Although participants were verbally encouraged to participate, the positioning of texts and the spatial configurations of the room discouraged input. However, participants critiqued the delivery and the micro-environment of the meetings through a demonstration of their self-worth and citizenship.

DECORUM: WHY "STUFF" MATTERS

Throughout the public input document, citizens—advisory and focus group members—reminded planners that environment, language, history, customs, and precedents mattered and existed. Continually, they insisted on specifying existing environmental conditions, appropriate references to history, citations of previous land-use planning efforts, and correct pronunciations. These "details," as one planner put it, did not always necessarily jibe with the rhetorical goals of the presenters (cast a vision), but it is this insistence on the appropriate (that which fits the occasion, message and the audience, and in this case, the environment) that allowed participants to influence the drafting process (Poulakos 41).

In a particularly telling incident of the appropriate, citizens insisted on correct pronunciation. As reported, participants did not ask questions during the PowerPoint presentation. Questions were sparse in the advisory committee meeting until the male presenter mispronounced Mesilla, using the short "I" sound instead of the long Spanish "e." A participant corrected him, he responded, "Okay, okay, okay," waved his hand, and continued with his explanation of his methods. The mispronunciation might have worked to remind participants that presenter was not from the community, thus his ideas, and consequently the maps, might not be appropriate to the area. Whatever thoughts the mispronunciation triggered, questioning and commenting increased after the mispronunciation. Interestingly, at the citizen's meeting an awkward pronunciation of "pēcān" also led to an increase in questions and comments.

Pronunciation mistakes appeared to open a space for participants to begin to assert authority, not just over proper pronunciations, but also over the planning process. For example, the advisory committee members began to point out a few existing neighborhoods and a planned school were missing from the maps. One participant said, "How can you predict growth when you do not accurately record existing conditions?" (field notes). They transferred their local knowledge of the community to interpret the maps and make corrections. Furthermore, citizen meeting participants began questioning the worth of creating the master document in light of the history of planning in the county and city. One participant cited the history of planning in the county as a reason to question the Alternative Land Use map: "For thirty years, people have been coming here and telling us to plan in those node circles. It has never happened. We need to plan for flexibility because we do not know for sure that these growth predictions will come true" (field notes). From the opening of a mistake in pronunciation, participants in both meetings started to reposition themselves as experts, as those who know what is appropriate.

John Poulakos ties a kairotic, situated rhetoric to the appropriate thusly: "Rhetoric is the art which seeks to capture in opportune moments that which is appropriate and attempts to suggest that which is possible" (36). Poulakos seeks to delineate how rhetoric might best operate to impact a given moment or process in a particular place. Furthermore, Poulakos's definition embraces decorum, or appropriateness, an ancient idea derived from Cicero's translation of the Greek *to prepon*, an idea that allows for the complexity of location. Poulakos explains the nature of appropriateness thusly:

> When appropriate, speech is perfectly compatible with the audience and the occasion it affirms and simultaneously seeks to alter. An appropriate expression reveals the rhetor's rhetorical readiness and evokes the audience's gratitude; conversely, an inappropriate expression indicates a misreading on the rhetor's part and a mismeeting between rhetor and audience. If what is spoken is the result of a misreading on the part of the rhetor, it subsequently becomes obvious to us, even to him, that "this was not the right thing to say." (41)

In contemporary studies of rhetoric, appropriateness or decorum has been negatively cast as following the manners of the ruling class or kowtowing for favor with the audience, but Poulakos's definition can be extended for understanding the relationship between rhetoric and location. The notion of compatibility among the message, occasion and audience asks the rhetor, not to simply court the audience's favor, but to properly understand, and respond to, the ecology of the situation, including the language, history, customs, and precedents of the current rhetorical situation.

Thus, participants in both meetings drew on their knowledge of place and what was considered appropriate to the place to reposition themselves as more than receivers of expert advice. They understood the ecology of the situation and used that understanding to question the rhetors' proposal. Their experiences as members of the community and their respective ties to the community allowed them to ask pertinent questions and reposition themselves as authorities. For example, a resident of a colonia

cited the history of unregulated planning in her neighborhood before asking how the Alternative Land Use map addressed existing conditions. A
firefighter prefaced his question about adequate fire and emergency infrastructure by explaining how fire services are presently funded. A farmer's
wife eloquently explained the relationship between land and wealth in
agriculture communities in response to the presenter's claim that explaining what the presenters meant when they said the plan would protect
agriculture was not relevant at the moment. A New Mexico State University alumnus explained the importance of maintaining ties between
the university and the community before asking how the maps accounted
for the traditions of New Mexico State University. The knowledge of the
appropriate within the context of the region (the history of the colonias,
the relationships between land and wealth, the university and the region)
allowed them to question how the process served and accounted for the
particular needs of the community.

OF CONFLICT, LITERACY, AND CIVIC PARTICIPATION

Experts, the New York based planners, came to town, and they are still
coming to town, trying to build consensus around a much debated planning document. At the time of this writing, *Vision 2040* has still not
been adopted as a master planning document by the major stakeholders
(the county commission, town, city, and village councils). As this familiar
story unfolded, two things became clear. First, text alone cannot negotiate
differing visions. The differing visions shared in the advisory community
and the focus groups needed to be negotiated and aired by participants—
those who have a stake in producing Hauser's active society marked by
conflict. Second, what the planners and presenters saw as details (the
complexities of location) were actually the foundations upon which the
success or failure of the planning process pivoted. Because rhetoric is rela-
tional and relationships are often based on shared knowledge and experience, the seemingly minute details of addressing wording differences or
the open then quickly closed mouth of an audience member, the seat-
ing arrangements in a room, or local word pronunciation, can either lay
waste to opportune moments and the possible or capture them just in
time. Most importantly, these details of lived experience, delivery and
decorum were tied to the particulars of the region, the very subject which
should interest those wanting to shape the lived environment. For community literacy and environmental advocates, this means three things: 1)
we need to see the conflicts created by different experiences of place as a
generative producer of society, capable of creating processes build around
"relationships, conversation, and in-depth choices"; 2) local knowledge,
arising from the lived environment, must be as respected as expert knowledge in a specialized field; and 3) to encourage civic participation around

environmental discourse, we may do well to foster an awareness of the particulars of place. For in Doña Ana County, literacies sponsored by living in, and understanding a locality, were deployed to rhetorically intervene in political processes.

WORKS CITED

Adams, Catherine. "PowerPoint Habits of Mind and Classroom Culture." *Journal of Curriculum Studies* 38.4 (2006): 389–341. Print.

Burke, Kenneth. "Definitions of Rhetoric." Web. 3 Apr. 2012. http://www.stanford.edu/dept/english/courses/sites/lunsford/pages/defs.htm

Clapp, Donna. "Las Cruces, New Mexico Metro Spotlight." *Business Facilities*, June 2007: 36–37. Print.

Goggin, Peter, and Elenore Long. "The Co-Construction of a Local Public Environmental Discourse: Letters to the Editor, Bermuda's *Royal Gazette*, and the Southlands Hotel Development Controversy." *Community Literacy Journal* 4.1 (Fall 2009): 5–26. Print.

Habermas, Jurgen. *The Structural Transformation of the Public Sphere.* Cambridge, MA: MIT P, 1991. Print.

Hauser, Gerald. "Rethinking Deliberative Democracy: Rhetoric, Power and Civil Society." *Rhetoric and Democracy: Pedagogical and Political Practices.* Ed. Todd F. McDorman and David M. Timmerman. East Lansing: Michigan State UP, 2008. 225–264. Print.

Knaap, Gerrit-Jan. "A Requiem for Smart Growth." Presented at "Planning Reform in the New Century," Washington University Law School, St. Louis, MO, December 2004. Web. 3 Apr. 2012.

Lindgren, Tim. "Composition and the Rhetoric of Eco-Effective Design." *Coming Into Contact: Explorations into Ecocritical Theory and Practice.* Ed. Annie Merrill Ingram, Ian Marshall, Daniel J. Philippon, and Adam W. Sweeting. Athens: U of Georgia P, 2007. Print.

Mason, Eric. "Greening the Globe, One Map at a Time." *Community Literacy Journal* 4.1 (Fall 2009). 93–104. Print.

Mesilla Valley Economic Development Alliance. 21 Oct. 2008. MVEDA. Web. 25 Nov. 2011.

Poulakos, John. "Toward a Sophistic Definition of Rhetoric." *Philosophy and Rhetoric* 16 (1973): 35–48. Rpt. in *Contemporary Rhetorical Theory.* Ed. John Louis Lucaites, Celeste Michelle Condit, and Sally Caudill. New York: Guilford P, 1999. 25–34. Print.

Prior, Paul, Janine Solberg, Patrick W. Berry, Hannah Bellwoar, Bill Chewning, Karen Lunsford, Liz Rohan, Kevin Roozen, Mary Sheridan-Rabideau, Jody Shipka, Derek Van Ittersum, and Joyce Walker. "Re-Situating and Re-Mediating the Canons: A Cultural-Historical Remapping of Rhetorical Activity." *Kairos: A Journal of Rhetoric, Technology, and Pedagogy* 11.3 (2007). Web. 3 Apr. 2012.

Sartori, Jason, Terry Moore, and Gerrit Knaap. "Indicators of Smart Growth in Maryland." The National Center for Smart Growth Research and Education. January 2011. Web. 3 Apr. 2012.

Schroeder, Christopher. "Notes Toward a Dynamic Theory of Literacy." *The Locations of Composition.* Ed. Christopher J. Keller and Christian R. Weisser. Albany: State U of New York P, 2007. 267–88. Print.

Soja, Edward. *Seeking Spatial Justice.* Minneapolis: U of Minnesota P, 2010. Print.

Spirn, Anne Whiston. "Restoring Restoring Mill Creek: Landscape Literacy, Environmental Justice and City Planning and Design." *Landscape Research* 30.3 (2005): 395–413. Print.

U.S. Bureau of the Census. *County and City Data Book*. Washington, DC: U.S. Government Printing Office, 2000. Print.

Vision 2040. City of Las Cruces and Doña Ana County. Web. 27 Nov. 2011. http://vision2040.nmsu.edu/index.html

14 Place-Identity and the Socio-Spatial Environment

Rick Carpenter

"Environmentalism" is a way of living in the world without forgetting how we come to know and value that world. (Maxcy)

In his review of environmental studies within the field of professional communication, M. Jimmie Killingsworth argues that pervasive hyper-specialization, a trend first established by Cold War–era environmental rhetoric and eco-criticism, probably lessened the potential impact and influence of these studies by unnecessarily limiting their readership. The mistake, Killingsworth laments, was in thinking that "only certain groups are touched by ecological concerns and the interest in place" (361). The unfortunate result of such thinking has been a tendency to construe nature as simply a source of problems for experts to address. As a corrective, he proposes a revised pedagogy and research program, one informed by the insights and methodologies of eco-composition and eco-poetics. Stressing place-consciousness and localization, such a program would be sensitive to a wider range of concerns and issues. Although Killingsworth's specific focus is professional communication, his critique touches upon and is relevant to studies of environmental discourse in general. Scholars in environmental rhetoric are now beginning to answer Killingsworth's call for studies that begin with the question of place. The present collection represents an important contribution to this enterprise.

Importantly, as Killingsworth's reading of eco-composition makes clear, investigations should start not with place but "with the *question* of place" (365; emphasis mine). As Dolores Hayden reminds us, "Place is one of the trickiest words in the English language, a suitcase so overfilled one can never shut the lid" (15). As over-determined signifier, *place* swirls about and between us, usually unnoticed or unacknowledged: *Place it over there. She was in a really bad place. C'mon over to my place. Know your place. Rivers are important places. There's no other place I'd rather be.* And then there are questions of scale and delineation: How does one define, locate, and frame place in space? Where does one place stop and another begin?

This chapter seeks to explore the question of place in relation to the environment through a case study analysis of the rhetoric that characterized the debate over a proposed biomass electric generating plant in Valdosta, Georgia. Specifically, I examine the crucial role of space and place in how the conflict was framed by the opposing sides and received by diverse publics. Despite the polarizing discourse of the various stakeholders that often oversimplified issues, the debate did not ultimately stall in a paralyzing impasse. As this chapter will show, opponents of the biomass plant eventually succeeded in gaining widespread public support for their cause by employing a geographic rhetoric that fostered coalition-building through the socio-discursive production of space as more inclusive place. As a consequence, this study raises important questions about the relationship between politics of space and structures of power, and the importance of place to collective mobilization.

THE BIOMASS CONTROVERSY IN VALDOSTA

Toward the end of 2007, the Valdosta-Lowndes County Industrial Authority (the Authority), an organization charged with encouraging and overseeing industrial and economic development in Valdosta and Lowndes County, Georgia, began talks with Wiregrass Power, LLC, a subsidiary of Sterling Energy Assets, a Georgia-based company, in hopes of attracting a biomass electric generating facility to the area. Biomass plants generate electricity by burning organic matter, usually wood by-products or residue gathered from factories and farms. The plan was announced to the public a few months later, with a public hearing held to provide local citizens with more details about the proposed plant. A month later, the County Board of Commissioners unanimously approved the rezoning of the proposed site, allowing the Authority, who had previously purchased the land, to lease the property to Wiregrass. From the beginning, Dr. Michael Noll, a professor of geosciences at Valdosta State University (VSU), expressed his concerns about the health risks associated with biomass plants in a number of public forums and media. Over the following months, an increasing number of local residents also began to voice concerns regarding air quality and health risks, prompting the mayor to acknowledge their concerns in his annual State of the City address. Soon after, the Georgia Environmental Protection Division (EPD) approved Wiregrass Power's application for an Air Quality Operating permit to construct and operate the biomass plant. In response, a number of local organizations became actively involved in the controversy, notably Wiregrass Activists for Clean Energy (WACE), a group cofounded by Noll; Students Against Violating the Environment (SAVE), a student organization at VSU; and the local branch of the NAACP. As protesters became more and more vocal, relations between members of the Authority and activist leaders grew increasingly strained and tense.

As is often the case in environmental debates, stakeholders soon aligned in terms of a supporter and opponent binary. The discourse of the controversy, unsurprisingly, frequently took the form of what Killingsworth and Palmer term *ecospeak*, with public debate reduced to two opposing options: pro-biomass or anti-biomass. However, despite the strict lines of demarcation between the two positions, the conflict cannot be simplistically classified according to the traditional oppositional dichotomies that have characterized much of American environmental rhetoric. In this regard, the biomass debate in Valdosta echoed the controversy over the Cape Wind project, which involved a field of offshore wind turbines approved for Nantucket Sound. Kimberly Moekle notes that the Cape Wind controversy was complicated by the fact that the rhetoric of supporters and opponents often blurred the boundaries between such historical polarizations as utilitarian/ romantic, conservationist/preservationist, and environmental/developmental. For instance, as a business venture that would reduce carbon dioxide emissions (among other benefits), the Cape Wind project, Moekle observes, represents both utilitarian and environmental viewpoints.

In a similar fashion, advocates and opponents of the proposed biomass plant in Valdosta drew upon the same traditional rhetorical modes in their discourse about place: environmental, developmental, and local. Proponents argued that the biomass plant represented a safe, "green" source of renewable energy that would bolster the local economy by creating new jobs and saving taxpayer money. Meanwhile, opponents maintained that the biomass plant would release dangerous particles into the air and that, given the local geography (Lowndes County borders Florida, the "Sunshine State"), city and county governments should instead pursue solar power alternatives. This intermingling of rhetorics supports Laura Johnson's assertion that no single rhetoric is adequate in the contemporary climate of environmental politics, only a mixture of rhetorics (44). Given that both sides spoke of place and the environment in such similar ways, it is little wonder that the public, at least in the beginning, was largely confused, uncertain, or unaware.

The primary concern of the opponents—air quality—was also their most salient argument; however, by itself, when too far removed from the exigencies of the local, it was not necessarily their most persuasive. In relying primarily upon a scientific mode of discourse, opponents faced the difficult task of establishing their credibility and right to be heard—*ethos*—when decision makers repeatedly labeled them as a "'fringe' group" ("Thumbs") and they lacked the authority of medical credentials. As Derek Ross notes, "to effect change, one must convey expertise and authority, even if one has not dedicated their life to the study of a particular issue." The strategy opponents most commonly employed was to present a wealth of specific, sometimes highly technical information drawn from sources that readily conveyed scientific expertise, such as *The American Journal of Public Health* and the American Lung Association. For example, WACE provides a document on their website, "Myths Versus Facts About the Proposed Biomass Plant," that, among other things, asserts that biomass is less clean than coal. As evidence, the document lists the future emissions of the proposed plant in Valdosta: "PM 0.025 lb / mmbtu, SO2 0.090 lb / mmbtu, NOx 0.090 lb / mmbtu." Opponents who spoke at government meetings would often present the same medical source evidence each month, eventually causing both sides to express frustration.

Proponents countered by holding public forums in spacious public buildings and bringing in an environmental consultant, Robert McCann, Jr. from Golder Associates, Inc., to work on the project and allay community fears. The scientific discourse of the proponents was also almost always intermixed with regulatory discourse. Rather than directly question or counter the presented medical evidence (which they were likely unable to do), they emphasized that the plant had been granted an air permit by the EPD, and that the plant's emissions would be well below the maximum allowed by the permit. Implied in this assertion is that the EPD, as a regulatory body charged with protecting the state's air quality, would have the resources and thus the (scientific) knowledge to know whether biomass is a safe energy source. In a community as historically conservative as Valdosta, official discourse, as a rule, is more readily accepted—and less likely to be questioned—than the discourse of "fringe" activists.

Yet opponents did eventually manage to build a wide base of public support from across the community. How were they able to effectively mobilize a community-wide coalition of diverse participants? From a rhetorical perspective, scientific and economic facts alone do not adequately account for the opponents' persuasive success. Nor do emotional appeals. A close examination of the competing rhetorics reveals key differences in how the opposing sides framed the issues most important to their own causes, even when those issues (such as sustainability) were identical or nearly so. I propose that although both sides framed the debate in terms of the local, the opponents strategically employed the rhetoric of place in order to create and sustain a community identity situated firmly within—and indeed embodied by—the local landscape, a geographic rhetoric that proved effective in linking together diverse publics.

GEOGRAPHIC RHETORIC AND POLITICS OF SPACE

Given that the Authority's sole reason for being, its *raison d'être*, is to develop the local economy, it is unsurprising that the proponents worked to situate the biomass plant firmly within the spatiality of the local. For instance, the Authority downplayed Wiregrass Power's financial interests while highlighting local economic benefits. And Gilbert Waldman, vice president and general manager of Sterling Energy, stated at a public meeting, "There will be nothing to impact the local area" (Pinholster "Biomass"). Proponents were also quite geographically specific when discussing the location chosen for the plant. When asked why the particular site had been chosen, Authority Project Manager Allen Ricketts answered that the "proximity to reclaimed water, the access to significant transportation arteries and main power grid lines were factors" (Pinholster "Public").

When Waldman was later asked the same question, he described the site as being an "ideal location from an electricity standpoint" and then listed similar reasons (Ramos). At another time, Waldman proclaimed that the company's plan as owner and operator of the plant was to become "a part of the community" (Pinholster "Biomass").

Rhetorically, this is the conventional manner of writing and speaking about the geography of place. It is the rhetoric of boundaries, borders, and margins, of movement through space and habitation in place. The environment is something to be surveyed, mapped, and represented. Nature is a location *out there*, somewhere to visit, something to see. This is the rhetoric of delineation and difference, of familiarity and fear, of organization and arrangement. Within this frame, scientific and utilitarian rhetorics speak of nature as object and resource, respectively, and the built environment becomes the human sphere, separate from the natural environment. Accordingly, the discourse of the proponents assumes a unity of place for the local community, a human solidarity in/of the built environment (even when the construction is simply a dirt road). Valdosta from this perspective is a place on the map, a geometric ordering of streets, and nature is but a standing reserve, to use Heidegger's term. The community exploits this valuable resource and, importantly, is defined—located—to no small extent by that exploitation. The drawback of this particular mode of geographic rhetoric is that it tends to ignore difference and elide diversity; a particular actor is either local or not. In assuming an unproblematic, uncomplicated common identity for their audience, rhetors can neglect to foster identification, which, as Kenneth Burke asserts, is always a rhetorical activity.

The problem is that traditional geographic rhetoric pretends to arhetoricity. The proponents' rhetoric of place was dependent on a Euclidean notion of space as neutral and fixed. Space in this view is transparent (Blunt and Rose) and absolute (Smith and Katz), and thus unquestioned. As a result, the socio-discursive construction of space and place is elided. In this manner, the relationship between imagined geographies and material conditions, what Nedra Reynolds terms the *politics of space*, including the use of space to construct and normalize unequal power relations, is masked or denied. Instead, space is treated as abstract. According to influential spatial theorist Henri Lefebvre, the spatial economy of abstract space "valorizes certain relationships between people in particular places . . . and thus gives rise to connotative discourses concerning those places; these in turn generate 'consensus' or conventions" (56). The spatial practices that (re) produce structures of power are naturalized (Harvey), and people learn to know their place (in both senses of the term) in material-metaphoric space. Within the dominant socio-spatial order are clear lines of distinction and difference. You either belong or you don't; if you're not an insider, you're an outsider. By a similar process, humans are separated from nature. The natural environment becomes the *out there*, a different place in the same local space.

The ideology of transparent, abstract space undergirds all aspects of the proponents' rhetoric. On one hand, their discourse seeks to locate the benefits of the biomass plant firmly within the borders of the local. On the other hand, it constructs the local place as homogeneous, a primary objective of abstract space. There are simply "customers" (in the social/built environment) and "a variety of local sources" (in the natural environment) (Pinholster "Public"). And it is "Georgia" that "is well positioned," as a "greater market for landowners" is framed as an environmental benefit ("Our Opinion"). Everyone is equal, and everyone gains.

Except, of course, that everyone isn't and everyone doesn't. Even though proponents expended considerable energy and expense to allay community concerns about the potential environmental impact of the biomass plant, their discourse would inevitably construe their audience in narrow, exclusionary ways. For instance, proponents pointed out that the plant had been granted an air permit by the state EPD; later, however, after activists continued to cite air quality concerns, Ricketts remarked, "'With regard to citizen concerns about air quality emissions or other concerns that they have raised, it would not have an impact on the agreement [with Wiregrass Power], because the design of the facility is governed by the Air Quality Permit'" (Rodock "Biomass"). This statement seems to suggest that citizens' concerns no longer matter. But to whom? These concerns still certainly mattered to numerous interested parties. The unspoken audience, those whose fears Ricketts appears to be trying to assuage, is the proponents themselves, nearly all of whom were wealthy business leaders, many with direct financial stakes in the project and none with residences in the immediate vicinity of the proposed site of the plant. It was almost as if Ricketts proclaimed, *We no longer care because* we *have the legal authorization and there's nothing* they *can do about it.* Waldman was also rather direct when he remarked, "I'm not interested in being environmentally altruistic; it is purely financial to stay within air permit requirements" (Pinholster "Biomass"). Opponents sometimes pointed out loopholes in the air permit or remarked upon the political motives guiding appointments to the EPD; proponents never addressed those issues.

Similarly, proponents generally ignored remarks made by opponents concerning issues of environmental injustice and racism. In the discourse of the proponents, the plant would be located simply (and unproblematically) in "the county." The only acknowledgement by proponents of the politics of space came after the Valdosta-Lowndes County branch of the NAACP issued a public statement claiming the location of the plant as a form of environmental racism.[1] In response, Authority Executive Director Brad Lofton stated that the Authority had requested a one-mile demographic analysis of the plant and the results, which were based on census data from 2000, indicated that the area was predominantly white. This data, however, failed to recognize changes that had occurred since 2000, including the construction of two schools and a residential area that were all predominantly black. Nor did it consider the seven predominantly black churches located within a two-mile radius of the proposed plant (Pinholster "Growing"). Here, Lofton's scientific rhetoric is framed by the geographic but in a way that fails to recognize place as lived experience; his data is abstracted from the very environment, and thus the individuals and groups, it claims to represent.

"BIOMASS? NO!": SPATIALITY AND THE LIVED ENVIRONMENT

Opponents' use of geographic rhetoric, in contrast to that of the proponents', was characterized by a high degree of specificity and attention to distinction and detail. Whereas for proponents the plant would be close to a source of reclaimed water, for opponents the plant would be "at the intersection of Perimeter Road and the Statenville Highway" (Condrey). When Noll spoke at a meeting early in the conflict, he inquired as to "whether the diversion of the outflow from the plant and adjacent Mud Creek Wastewater Treatment plant . . . [would] negatively affect downstream flows in the Mud Creek/Alapaha River system" (Fulton). What's more, geographic rhetoric featured more prominently and explicitly in the opponents' discourse than it did in the proponents'. Of note is the fact that the proponents had to be asked to explain their choice of location. Conversely, the first openly negative statement about the plant that appeared in the *Valdosta Daily Times* (*VDT*), an anonymous comment in the "Rant & Rave" section, expressed explicitly geographical concerns:

> Here we go again. Looks like us poor folks on the south end of Lowndes County will get the shaft once more. I am referring to the biomass plant they are going to build. Yes, we are going to get the smoke, stench, lower price of our property. I wonder why this type of thing is never built near a county commissioner's home. That big pecan orchard will make an ideal location. Oh, by the way, remember they wanted to build a hotel near Wild Adventures. It was unanimously denied. Wonder why? ("Rant, June 18")

Crucially, as this "rant" illustrates, the geographic rhetoric of the opponents' discourse was characterized by a greater sensitivity to the embodied spatiality of the lived environment—the historical experiences, cultural backgrounds, and social positions of those who inhabit, embody, and, indeed, construct particular locations. This sensitivity became even more evident after the local branch of the NAACP issued its proclamation of environmental racism. In addition, opponents, as is common in public protests, took their message literally to the streets—staging marches, handing out flyers, and posting "Biomass? No!" signs throughout the city (see Figure 14.1). In doing so, they concretized the conflict by situating it materially in the spaces of everyday life, highlighting in the process the fact that space itself can be contested and, perhaps, re-imagined.

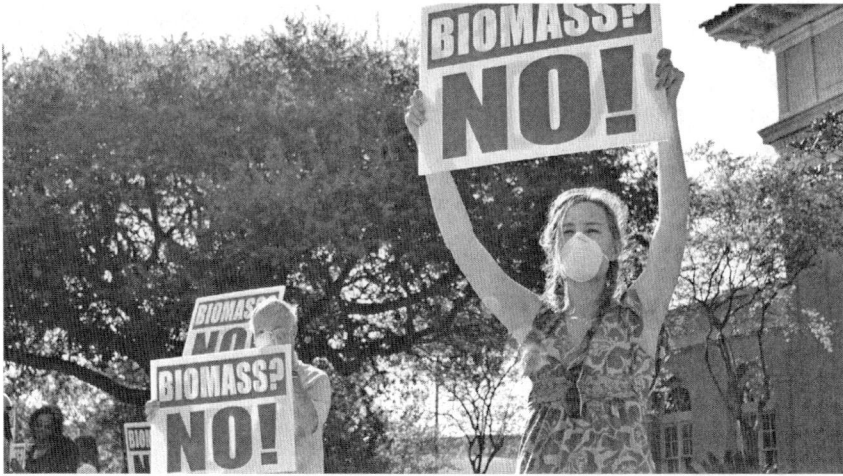

Figure 14.1   Protestors in downtown Valdosta. (Photo courtesy of *The Valdosta State University Spectator.*)

Such sensitivity to the materiality of difference is central to the enterprise of constructing an emergent social formation capable of motivating collective action for change. Resisting the hegemonic ideology of transparent, abstract space, which denies difference even as it works to maintain structures of power and privilege, was arguably the main problem faced by the opposition movement. As Pierre Bourdieu asserts, "We are constantly invaded and assaulted by the dominant discourse. A vast majority of journalists are often unconsciously complicit in the process, and it is incredibly difficult to break down that illusion of unanimity" (27). Illusions of unanimity and homogeneity support each other, so that opposing the dominant discourse risks being positioned as a fringe group. What's more, unless the illusion is broken, opposing arguments are, at best, likely to be given equal weight, leading to ambivalence inimical to a challenge to the status quo. Citing an American Lung Association statement about the health risks biomass incinerators pose for people with emphysema, asthma, diabetes, and heart disease at a public hearing with the EPD (Noll) may be dismissed if the audience doesn't know who to believe. Similarly, a charge of racism— environmental or otherwise—may lack persuasive power if perceived to be arbitrary (which is to say, misplaced or unplaced), especially in a region still recovering from a long history of racist policies and racial strife.

Although the biomass opponents worked tirelessly to change the minds of policymakers, they had little chance of success without a broad base of community support. The approach taken by governmental agencies in Valdosta and Lowndes County to public participation in decision making mirrors what Craig Waddell terms the "one-way Jeffersonian model": the public has a right to participate, but "it should be empowered to do so, simply and unproblematically, through a one-way transfer of expert knowledge" (142).[2] The opponents' success in strengthening the community's use of geographic rhetoric to construct a new social imaginary through evocation of a place-identity tied explicitly to the local-material environment, built and natural. The result was a sense of place that was more inclusive, not by emphasizing similitude but rather by calling attention to the socio-spatial production of difference. In other words, opponents employed geographic rhetoric as a sort of master frame that facilitated the development of an ecological perspective that in turn fostered the growth of the oppositional movement as localized place-identity.

In her influential essay "The Ecology of Writing," Marilyn Cooper proposes an ecological model "whose fundamental tenet is that writing is an activity through which a person is continually engaged with a variety of socially constituted systems" (367). Because we all move through and participate in multiple discursive ecosystems, the complexities of place remain even at the spatial scale of the local. Although Cooper never actually uses the term *environment*, her conception of systems easily accommodates its inclusion. From an ecological perspective, localization is an approach to place that is attentive to the particular discursive ecosystems inhabited or colonized by particular actors. To be local is to have a *sense of place*.

Highlighted in a localized approach to place is the notion of space and environment as process, as active agents in the construction of meaning. Cooper posits that ecological systems are "inherently dynamic" in that "they are made and remade by writers in the act of writing" (368). An ecological model reminds us that the relationship between discourse and environment is dynamic and reciprocal: environments shape writers but writers also shape their environments. As a result, writers' surroundings "are dynamic, difficult to define, and susceptible to the forces imposed by writers" (Dobrin and Weisser 568). In other words, environments, natural and constructed, are not stable constructs. They change. They can be created or destroyed. In examining ecologies of place and the environment, we need to account for change as well as stability in terms of both degradation and creation—including the production of place itself.

One of the key insights of human geography concerns the continual production of space (Mayer and Woodward 104). Rejecting the conventional view of spaces/places as the fixed, empty backgrounds to or containers of actors and actions, postmodern geographers such as David Harvey, Henri Lefebvre, Doreen Massey, and Edward Soja have reconceived places/spaces as dynamic scenes of action both constituted by and constitutive of the social. As Darin Payne reminds us, "Space creates frameworks for conception, action, and interactions; its design—whether natural or artificial—limits what we think and do, as well as with whom we do it" (484). In this sense, geography, as Reynolds asserts, is a "lived event" (10), for we compose our environments even as they are inscribed upon us. Produced by material and discursive processes, places are simultaneously physical and symbolic. Accordingly, as

Massey explains, places "are not so much bounded areas as open and porous networks of social relations" (121). As we move through and dwell in spaces (the stuff of material reality), we employ spatial metaphors and other spatial practices (such as writing) to shape and reshape space, the social lifeworlds in which we live. In essence, space is an abstract metaphor (although the spatial metaphors used to construct a particular space are frequently concrete) and imagined geography; a social construct formed through purposeful (inter) actions with/in the materially real by means of spatial practices; and a "realm of practices" (Reynolds 181) that structures our ways of knowing, seeing, and even being-in-the-world.

Accordingly, the self is also implicated in these socio-spatial practices. "[Q]uestions of 'who we are' are often intimately related to questions of 'where we are'" (Dixon and Durrheim 27). Exploring the relevance of the environmental psychological concept of place-identity to social psychology, John Dixon and Kevin Durrheim argue for a discursive approach to locating subjectivity. Such a perspective recognizes that identities are both located and socially produced. Applying this perspective to a study of national identity, Jodi Wallwork and John Dixon conclude that the rhetoric of nationalism is so effective because "nations" are imagined "not only as social categories" but also as "entities possessing a geographic and historical 'reality' that some-how exceeds their human membership" (22). Places are, by definition, more rigidly conceived—and thus experienced—than spaces, and the nation-state is arguably one of the most influential of places. As rhetorical construct, the "imagined community" (Anderson) of the nation-state is a mechanism for achieving stability as well as unity through the construction of a monolithic social space, a common sense of place and thus of self.

The conflict over the biomass plant in Valdosta demonstrates that the discursive practices and effects of place-identity also occur at spatial scales much smaller than that of the nation-state. The *idea* of Valdosta, like all environments is a socio-discursive construct; it is, in fact, the effect of a multiplicity of spaces and overlapping, often contested places: Valdosta State University, Moody Air Force Base, downtown, industrial district, commercial area near Interstate 75, country club, rural farms, "good" and "bad" neighborhoods, woodlands and rivers. As such, what Valdosta *means* is neither uniform nor static. Tim Cresswell argues that spaces and places are texts to be read, and, as such, "can be read in multiple and perhaps contradictory ways based on how readers are positioned ideologically to favor certain impressions over others" (Fosen 181). How one experiences a particular street, neighborhood, or business depends to no small degree on one's social location. In turn, streets, neighborhoods, and businesses come to not only signify but also produce and maintain particular identities.

The rhetoric of the controversy brought the pattern of place-based social relations in Valdosta to the fore of the debate. For some, as mentioned, the location of the plant was a form of environmental racism. Others emphasized simply that the plant would threaten the air quality in "Valdosta's poorest neighborhoods" (Gunning) or simply in the southern, more rural portion of the county. Still others *placed* the plant near elementary schools (Condrey). Michael Noll and other members of WACE, who typically drew upon science-based evidence from a variety of sources, also frequently included localized arguments. Writing about the high volume of water required by biomass plants, Noll remarked, "when you live in a region where water scarcity has become more common and water fights between Georgia and its neighboring states more intense over the years, such water dependent energy production makes no sense." Later in the controversy, opponents began to combine these place-based appeals, as evident in the following remarks by the president of SAVE at VSU:

> Notice how then old and young are impacted first because their immune systems are generally weaker. Well, this plant is in a predominantly black neighborhood on the Southside of Valdosta right down the street from two elementary schools, a middle school, and an elderly care center.
>
> So why should you care? How many of you will be here in two years? Three? For the next four years? How many of you have siblings that want to go to VSU? How many of you like playing outside? (Hurley)

Similarly, Leigh Touchton, president of the local branch of the NAACP, sought to broaden potential audience identifications by remarking, "'When a nuisance comes into the community, property values drop'" (Rodock "Lowndes").

As the public debate progressed and became more and more vitriolic, the rhetoric of the opposition became increasingly characterized by recognition of the multiple, at times conflicting, ways in which the controversy was situated (in place) and, thus, understood. However, rather than splintering the opposition into separate, competing movements by deepening already entrenched divisions in the community, this recognition actually helped to unify the opposition and thereby serve as the foundation for broad-based collective action, at least for the duration of the controversy and within the framework of the supporter/opponent binary. Because the conflict centered on environmental issues, new rhetorical spaces of identification could be conceived. This could occur as a consequence of the traditional conception of the environment as somehow outside the human world. A space of ambiguity, a kind of no-place, the environment could come to function within the discourse of the opponents as a source of shared locatedness that did not attempt to deny or erase the different motives, values, and interests of the various stakeholders. Indeed, their differences were both highlighted and placed. In this manner, the solipsism of the static spatial imaginary could be deconstructed and the localized place re-imagined as a complex, changing, interactive network of entangled spaces and places. Speaking of national identity, Wallwork and Dixon argue that "nationalist rhetoric works as a tool of mass mobilization not because all members of a nation share a homogeneous identity, derived from real or perceived similarities. It works because the category 'nation' can be considered flexibly and strategically in order to unite disparate groups" (22). The geographic rhetoric of the biomass opponents worked in much the same way. The highlighted *scene* of the socio-spatial interactions among the various stakeholders, the shared space and place-held-in-common, was the natural environment, from which they could construct a shared notion of locatedness that recognized yet overcame their heterogeneity.

In invoking place as source of identification and difference, the geographic rhetoric of the opponents called attention to rhetoricity itself, in particular the relationships among discourse, place, and power. By highlighting contradictory constructions of and experiences with/in place, the discourse of the opponents evoked the contradictions inherent in the social production of space, disrupting the ideology of transparent space. The absolute rationality of something such as census data, for instance, could be called into question. The discourse of the proponents, cast in sharp relief, was laid bare, its rhetoric of place shot through with relations of power. As a result, the divide between supporters and opponents was deepened even as the ties connecting the opponents (and constructing the newly imagined place-identity) were strengthened.

The discourse of the proponents was frequently dominated by the language of business: *customers, markets, members, companies,* and, perhaps most damning, *landowners.* In contrast and, arguably, in response, opponents made repeated and explicit references to structures of power and influence in the community, as the examples below illustrate:

> You are absolutely correct about the health hazards that the plant will produce. Unfortunately, knowing the past of the local city powers and . . . county higher-ups, good luck with them doing anything for the people and not just for themselves. ("Rant, Feb. 25")

> If the powers that be get their way, we will soon see the construction of a biomass incinerator in our city. (Noll)

> The biomass plant will be a major mistake for the area, burning trash, human waste, foliage and other matter as fuel. What goes in must come out people. Not to mention one of the pushers for this has family trying to supply some fuel for it. Not to mention the fact that nothing safe ever needs a buffer zone. Twenty-two acres purchased, eight acres used; the rest is a buffer zone, according to the meeting. ("Rant, Feb. 25")

> In recent weeks, as Valdosta's citizens begin to become more vocal, [County Chairman Ashley] Paulk and [Executive Director Brad] Lofton are now providing a narrative to the community that it is too late for anything to stop this project. In essence, the conservative Lowndes County government officials are telling the community "to get over it."

This arrogant response comes from someone who believes he and his collection of rich, conservative friends can get away with dumping this toxic, money pit into a majority-minority city, Valdosta, and the larger community of Lowndes County. (Davis)

When the companies are run out of one town or state, they move to places like Georgia, where environmental regulations are lax and businesses are given huge tax incentives to build these kinds of incinerators, and investors count on the local population being uninformed and apathetic. ("Valdosta NAACP")

The issue of investors became a major area of contention, mentioned frequently by opponents. When the *VDT* resorted to an Open Records Request to the Authority to obtain a list of investors, the Authority stated in the information provided that "a list of those investors is not available to the public" (Pinholster "Valdosta"). Indeed, the *VDT* published several reports of the Authority's reluctance to provide requested information. One article, "Valdosta Biomass Plant Fuels Questions," ends with this statement, "No information was provided by the Authority when asked by *The Times* for information regarding health or environmental hazards associated with the biomass plant" (Pinholster). Although the paper remained neutral or, at times, supportive of the plant at the beginning of the controversy, its official stance switched to adversarial/opposed as questions remained unanswered.

The animosity between the proponents and opponents reached a peak when the Lowndes County Board of Commissioners altered the established protocol of their meetings in a way widely perceived as intended to limit free speech and restrict public debate. Previously, during a designated Citizens-to-Be-Heard portion, any individual could speak freely about any topic. The approved amendment allowed the Board to impose restrictions on "'the time and manner of speech'" as well as "to limit speech" on subjects the board believed to be closed (Rodock "Lowndes"). The move became a public relations disaster for both the board and the Authority, who had already cut off communications with local media outlets (Rodock "Industrial").

The result was a collective mobilization the proponents had not anticipated. Echoing the opinion of many observers, Touchton remarked, "This was the most diverse group of people that I have ever seen for any citizen's group in Valdosta" (Rodock "Burning").

CONCLUSION: PLACE, AN ECOLOGICAL PERSPECTIVE

Opposition to the biomass incinerator plant in Valdosta grew from a small group of activists to a community-wide social movement. My contention is that the opposition coalition was not based solely or even primarily on politics, class, or race. Nor was it even wholly a product of environmental concerns or rhetorics. For much of the controversy, proponents were quite successful in countering opponent claims. Indeed, given the relative power and influence of the proponents in the community, one might have expected the controversy to be quickly forgotten by the local media and the majority of residents. The somewhat surprising success of the opponents in fostering widespread public support for their cause stems in large part from a geographic rhetoric that encouraged or even forced individuals and diverse publics to confront the multiple and sometimes conflicting constructions of place in the spaces of Valdosta. This recognition allowed for the creation of new spaces, entangled spatial networks that together comprised a re-imagined social imaginary vis-à-vis the natural environment. In other words, the opponents helped foster a more ecological perspective across a number of translocalized spaces, natural and built.

This ecological perspective contrasted sharply with the ideology of abstract, transparent space in the discourse of the proponents. Highlighted in the comparison were place-based mechanisms of social relations and power. The geographic rhetoric of the proponents argued implicitly for a particular socio-spatial organization, one that maintains the status quo. In this view, spaces and places, as empty backdrops to human action, are inconsequential, and environments are of little significance, except as resources to be exploited. At a public forum organized to reassure the community, Lofton reported that a study of the plant site had been conducted by Lovell Engineering. He added, however, that "neither the Authority nor Sterling were required to do an external study on the impact the biomass plant would have on surrounding environs" (Pinholster "Biomass"). This comment suggests a lack of regard for the environment as well as, crucially, a lack of understanding regarding how place constructions may discursively ground particular versions of identity (Wallwork and Dixon 26), and how local landscapes can come to embody common identifications. Throughout the controversy, the discourse of the proponents not only separated human lifeworlds from natural environments. It also construed these lifeworlds as somehow singular and uniform, and therefore devoid of material-social differences and inequalities. Their rhetoric emphasized localization while denying the locatedness of identity. As the public became increasingly aware of the place-identity relations within the community, the traditional rhetoric of localized placelessness was, for a growing number of people, both unpersuasive and insulting. In 2009, Lofton could confidently state that the plant represented "a desirable business opportunity for the community" (Harris). By 2011, many in the community were asking, *Opportunity for whom exactly? And at what cost?*

As the debate raged on, Wiregrass Power failed to meet the project goals established by the Economic Development Agreement (EDA) it had signed with the Authority. The EDA had set an April1, 2011 deadline for the successful completion of certain objectives related to the construction and operation of the plant (such as finalizing a power purchase/transmission agreement). The company was granted an extension to June 1 but again failed to meet the deadline. Authority Chairman Jerry Jennett reported that Wiregrass had not asked for an additional extension and that, even if they did, he was confident the request would be denied.

Just as activists were claiming victory, Sterling Energy, the parent company of Wiregrass Power, expressed its desire to purchase the land outright from the Authority, an option allowed by the contract it had signed with the Authority. At first, the Authority's reaction was to cite the legality of the matter, stating that they were powerless to stop the purchase, and, by extension, Wiregrass Power's right to do with the land as it wished. However, not long after, Valdosta City Manager Larry Hanson sent a letter to Sterling stating that the City would not provide the property with water and utility services. Because the biomass plant would require thousands of gallons of water an hour to operate, this effectively signaled the death knell for the plant. A month later, the Authority announced it had submitted a deal for Wiregrass to purchase the land with the condition that no biomass facility be developed there. And with that, the controversy came to a close.

Biomass as a source of energy has been criticized by a host of environmental groups. In locations across the United States as well as Europe, activists have organized in opposition to existing and proposed plants. Analyzing the rhetoric that characterized the debate over the biomass plant in Valdosta, however, reveals less about biomass as a possible source of energy than it reveals about the complex relationship between identity and environment, and the situated, discursive nature of that relationship. Examining how space is constructed as place through discourse can shed light on how place is construed in relation to the environment. The biomass conflict in Valdosta demonstrates that the manner in which places are conceived and represented can be reshaped and re-imagined in ways that make them both more equitable and more sustainable.

NOTES

1. A detailed discussion of environmental racism and justice is beyond the scope of this article. A useful place to begin such an examination is Dorceta E. Taylor's comprehensive "The Rise of the Environmental Justice Paradigm." See also Julia Miller Cantzler's study of the conflict between the Makah Indian Tribe and activists over hunting gray whales. Both studies also examine the concept of rhetorical framing in relation to social movements.

2. See Steven B. Katz and Carolyn R. Miller for a discussion of the difficulties and failures that typically characterize risk communication and public participation programs. Also, Emilian Geczi provides a useful discussion of the influence of market ideology and socioeconomic inequalities on public participation and the democratic process.

WORKS CITED

Anderson, Benedict. *Imagined Communities: Reflections on the Origin and Spread of Nationalism*. London: Verso, 1991. Print.

Bourdieu, Pierre. "A Literature from Below." *The Nation*, 3 July 2000: 25–28. Print.

Blunt, Alison, and Gillian Rose, eds. *Writing Women and Space: Colonial and Postcolonial Geographies*. New York: Guilford, 1994. Print.

Burke, Kenneth. *A Rhetoric of Motives*. Berkeley: U of California P, 1969. Print.

Cantzler, Julia Miller. "Environmental Justice and Social Power Rhetoric in the Moral Battle over Whaling." *Sociological Inquiry* 77.3 (2007): 483–512. Print.

Condrey, Jeffrey. "Health Dangers of Biomass Plant Are a Bigger Concern." Letter. *Valdosta Daily Times*, 4 Aug. 2010. Web. 5 Nov. 2011.

Cooper, Marilyn M. "The Ecology of Writing." *College English* 48.4 (1986): 364–375. Print.

Davis, Patrick. "Paulk–Perdue Partnership Played Central Role in Disputed Valdosta Biomass Plant." *Examiner.com*. Clarity Digital Group LLC, 26 Jan. 2011. Web. 7 Nov. 2011.

Dixon, John, and Kevin Durrheim. "Displacing Place-Identity: A Discursive Approach to Locating Self and Other." *British Journal of Social Psychology* 39 (2000): 27–44. Print.

Dobrin, Sidney I., and Christian R. Weisser. "Breaking Ground in Ecocomposition: Exploring Relationships between Discourse and Environment." *College English* 64.5 (2002): 566–589. Print.

Fosen, Chris. "Inside, Outside, Alongside: Geographies of a Writing Workshop." *Composing Other Spaces*. Ed. Douglas Reichert Powell and John Paul Tassoni. Cresskill, NJ: Hampton P, 2009. 163–184. Print.

Freudenberg, Nicholas, Manuel Pastor, and Barbara Israel. "Strengthening Community Capacity to Participate in Making Decisions to Reduce Disproportionate Environmental Exposures." *American Journal of Public Health* 101.s1 (2011): s123–s130. Print.

Fulton, Malynda. "County OKs Zoning for Power Plant." *Valdosta Daily Times*, 9 June 2009. Web. 5 Nov. 2011.

Geczi, Emilian. "Sustainability and Public Participation: Toward an Inclusive Model of Democracy." *Administrative Theory & Praxis* 29.3 (2007): 375–393. Print.

Gunning, Seth. Letter. *Valdosta Daily Times*, 27 Sept. 2009. Web. 5 Nov. 2011.

Harris, Kay. "Biofuels Are the 'New' Energy." *Valdosta Daily Times*, 10 May. 2009. Web. 5 Nov. 2011.

Harvey, David. "From Space to Place and Back Again: Reflections on the Condition of Postmodernity." *Mapping the Futures: Local Cultures, Global Change*. Ed. Jon Bird. Barry Curtis, Tim Putnam, and Lisa Tickner. New York: Routledge, 1993. 3–29. Print.

Hayden, Dolores. *The Power of Place: Urban Landscapes as Public History*. Cambridge, MA: MIT P, 1995. Print.

Hurley, Erin. Letter. *The Spectator*. Valdosta State University, 3 Mar. 2011. Web. 7 Nov. 2011.

Johnson, Laura. "(Environmental) Rhetorics of Tempered Apocalypticism in *An Inconvenient Truth*." *Rhetoric Review* 28.1 (2009): 29–46. Print.

Katz, Steven B., and Carolyn R. Miller. "The Love-Level Radioactive Waste Siting Controversy in North Carolina: Toward a Rhetorical Model of Risk Communication." *Green Culture: Environmental Rhetoric in Contemporary America*. Ed. Carl G. Herndl and Stuart C. Brown. Madison: U of Wisconsin P, 1996. 111–140. Print.

Killingsworth, M. Jimmie. "From Environmental Rhetoric to Ecocomposition and Ecopoetics: Finding a Place for Professional Communication." *Technical Communication Quarterly* 14.4 (2005): 359–373. Print.

Killingsworth, M. Jimmie, and Jacqueline S. Palmer. *Ecospeak: Rhetoric and Environmental Politics in America.* Carbondale: Southern Illinois UP, 1992. Print.

Lefebvre, Henri. *The Production of Space.* Trans. Donald Nicholson-Smith. Malden, MA: Blackwell, 1991. Print.

Massey, Doreen. *Space, Place, and Gender.* Minneapolis: U of Minnesota P, 1994. Print.

Maxcy, David J. "Meaning in Nature: Rhetoric, Phenomenology, and the Question of Environmental Value." *Philosophy and Rhetoric* 27.4 (1994): 330–346. Print.

Mayer, Danny, and Keith Woodward. "Composition and Felt Geographies." *Composing Other Spaces.* Ed. Douglas Reichert Powell and John Paul Tassoni. Cresskill, NJ: Hampton P, 2009. 103–120. Print.

Moekle, Kimberly. "The Vision or the View: Cape Wind and the Rhetoric of Sustainable Energy." *Rhetorics, Literacies, and Narratives of Sustainability.* Ed. Peter N. Goggin. New York: Routledge, 2009. 78–96. Print.

"Myths Versus Facts About the Proposed Biomass Plant." *Wiregrass-ace.org.* Wiregrass Activists for Clean Energy, 20 Feb. 2012. Web. 7 Nov. 2011.

Noll, Michael G. "Valdosta Gets Bio-Fooled." *Valdosta Daily Times,* 1 May 2010. Web. 5 Nov. 2011.

"Our Opinion: More Power on the Way." Editorial. *Valdosta Daily Times,* 11 June 2009. Web. 5 Nov. 2011.

Payne, Darin. "English Studies in Levittown: Rhetorics of Space and Technology in Course Management Software." *College English* 67.5 (2005): 483–507. Print.

Pinholster, Johnna. "Biomass Plant Said Good for Valdosta and Georgia." *Valdosta Daily Times,* 4 Nov. 2010. Web. 5 Nov. 2011.

———. "Growing Opposition of Proposed Biomass Energy Plant." *Valdosta Daily Times,* 27 Oct. 2010. Web. 5 Nov. 2011.

———. "Public Hearing Held for New Power Plant." *Valdosta Daily Times,* 14 May 2009. Web. 5 Nov. 2011.

———. "Valdosta Biomass Plant Fuels Questions." *Valdosta Daily Times,* 27 Oct. 2010. Web. 5 Nov. 2011.

Ramos, Kara. "Experts, Opponents Speak Out at Biomass Plant Meeting." *Valdosta Daily Times,* 9 Dec. 2010. Web. 5 Nov. 2011.

"Rant and Rave for February 25, 2010." Letter. *Valdosta Daily Times,* 25 Feb. 2010. Web. 5 Nov. 2011.

"Rant and Rave for June 18, 2009." Letter. *Valdosta Daily Times,* 18 June 2009. Web. 5 Nov. 2011.

Reynolds, Nedra. *Geographies of Writing: Inhabiting Places and Encountering Difference.* Carbondale: Southern Illinois UP, 2004. Print.

Rodock, David. "Biomass Plant Misses Deadlines." *Valdosta Daily Times,* 3 Apr. 2011. Web. 5 Nov. 2011.

———. "Burning Issue Put to Rest." *Valdosta Daily Times,* 1 June 2011. Web. 5 Nov. 2011.

———. "Industrial Authority Receives County Assessment." *Valdosta Daily Times,* 18 Oct. 2011. Web. 5 Nov. 2011.

———. "Lowndes Board of Commissioners OKs Amendment to 'Citizens to Be Heard.'" *Valdosta Daily Times,* 26 Jan. 2011. Web. 5 Nov. 2011.

Ross, Derek. "Course Review: Environmental Rhetoric, Ethics, and Policy—Teaching Engagement." *Present Tense: A Journal of Rhetoric in Society.* 2.1 (2011). Web. 15 Oct. 2011.

Smith, Neil, and Cindi Katz. "Grounding Metaphor: Towards a Spatialized Politics." *Place and the Politics of Identity*. Ed. Michael Keith and Steve Pile. New York: Routledge, 1993. 67–83. Print.

Taylor, Dorceta E. "The Rise of the Environmental Justice Paradigm: Injustice Framing and the Social Construction of Environmental Discourses." *American Behavioral Scientist* 43.4 (2000): 508–580. Print.

"Thumbs Up, Thumbs Down." Editorial. *Valdosta Daily Times*, 25 Mar. 2011. Web. 5 Nov. 2011.

Touchton, Leigh. "Environmental Racism." Letter. *Valdosta Daily Times*, 18 Aug. 2010. Web. 5 Nov. 2011.

"Valdosta NAACP Claims Environmental Racism." *Valdostanaacp.com*. Valdosta NAACP, n.d. Web. 7 Nov. 2011.

Waddell, Craig. "Saving the Great Lakes: Public Participation in Environmental Policy." *Green Culture: Environmental Rhetoric in Contemporary America*. Ed. Carl G. Herndl and Stuart C. Brown. Madison: U of Wisconsin P, 1996. 141–165. Print.

Wallwork, Jodi, and John A. Dixon. "Foxes, Green Fields and Britishness: On the Rhetorical Construction of Place and National Identity." *British Journal of Social Psychology* 43 (2004): 21–39. Print.

Afterword

Kim Donehower

> The most striking feature of contemporary moral utterance is that so much of it is used to express disagreements; and the most striking feature of the debates in which these disagreements are expressed is their interminable character. I do not mean by this just that such debates go on and on and on—although they do—but also that they apparently can find no terminus. There seems to be no rational way of securing moral agreement in our culture. (MacIntyre, *After Virtue* 6)

MacIntyre's observation seems to fit the central conflicts that run through many of the chapters in this collection: the "collision" Rebecca Powell describes in her chapter between environmental discourse and community literacy that result in land-use planning projects that "never reach fruition"; the poles of typical ecological debates that Rick Carpenter characterizes in his chapter as "utilitarian-romantic, conservationist-preservationist, and environmental-developmental." Whether the issue at hand is hydrofracking in the Marcellus Shale, mining copper in Superior, Arizona, or uranium at Mt. Taylor, New Mexico, the sides in many environmental controversies can be charted within the polarities Carpenter identifies.

In the paragraphs following the epigraph above, Alasdair MacIntyre gives examples of "well-known rival moral arguments" (7). His intent is to show that "we possess no rational way of weighing the claims of one as against another" (8). For example, MacIntyre describes three positions on the notion of "just war":

1 (a) A just war is one in which the good to be achieved outweighs the evils involved in waging the war and in which a clear distinction can be made between combatants . . . and innocent non-combatants. But in a modern war calculation of future escalation is never reliable and no practically applicable distinction between combatants and non-combatants can be made. Therefore no modern war can be a just war and we all now ought to be pacifists.

(b) If you wish for peace, prepare for war. The only way to achieve peace is to deter potential aggressors. Therefore you must built up your armaments and make it clear that going to war on any particular scale is not necessarily ruled out by your policies.

(c) Wars between the Great Powers are purely destructive; but wars waged to liberate oppressed groups, especially in the Third World, are a necessary and therefore justified means for destroying the exploitative domination which stands between mankind and happiness. (7)

MacIntyre points out that these arguments suffer from "conceptual incommensurability":

> Every one of the arguments is logically valid . . . the conclusions do indeed follow from the premises. But the rival premises are such that we possess no rational way of weighing the claims of one against another. For each premise employs some quite different normative or evaluative concept from the others, so that the claims made upon us are of quite different kinds. In [the just war argument] premises which invoke justice and innocence are at odds with premises which invoke success and survival. (8)

MacIntyre does not use ecological debates as examples, but we can see from the essays in this collection how such a debate might look in the terms of MacIntyre's analysis:

1. Depressed rural areas need well-paying jobs. Extract industries provide these jobs, often at much higher wages than other employers. Therefore, natural resources in sparsely populated areas should be tapped for both the good of the larger economy and to improve specific local rural economies.
2. Many places are linked to cultural and spiritual identity for specific groups. Destruction of these places for resource extraction damages or destroys cultural and spiritual identity. As part of a pluralistic society that respects groups' rights to cultural and spiritual identity, we must prevent the destruction of places that are linked to those identities.
3. Extract industries invariably cause ecological damage, affecting the health of people, livestock, and local flora and fauna. Therefore, extract should not be permitted.

Each of these arguments is internally consistent; as MacIntyre says, their conclusions proceed from their premises. But the "normative and evaluative concepts" underlying their premises stem from fundamentally irreconcilable values. Argument (a) sees economic benefit as the wellspring of other benefits; (b) privileges pluralism and meaning-making; (c) values health as the bedrock of well-being. Each of their conclusions proceed rationally from their underlying values, but how are we to question those underlying values and weigh them against one another to make practical decisions in individual cases?

As Brian D. Cope shows in Chapter 2, we can attack an opponent from within its own value system—demonstrating, for example, how mining companies create a "false exigency" within a rhetoric privileging job creation because those jobs are short-lived and unsustainable. Similarly, in Chapter 4, Sally Said describes environmental economist Thomas Power's argument that opposing environmental regulation for short-term economic benefit actually "undermin[es] the economic future of a region." These seem productive strategies, arguing from within an opponent's own unstated assumptions. But as we see in Chapter 7, attempting to persuade inside an opponent's framework, or with a set of common terms, is not always possible. In Chapter 7 Cynthia Haller offers an example of what happens when competing groups who try to advocate for their positions within a shared paradigm run up against their underlying conceptual incommensurability. She writes, "In some ways, this paradigm of sustainability has been productive for environmentally sound policy setting: its terms enable environmentalists and developmentalists to argue about their concerns in a shared language." But, Haller continues,

> the sustainability paradigm . . . is no panacea for environmental policy. Appeals to sustainability can sometimes obscure very real differences between policy positions competing for the validation of public audiences. Logical appeals to stewardship and sustainability, for instance, can be used to argue for conflicting positions in a discussion of whether and how to regulate agricultural use of fertilizers. On the one hand, fertilizers increase yield and thus minimize the need to develop land . . . On the other, runoff from fertilizer use endangers water resources.

For MacIntyre, this impasse is nearly impossible to resolve. He argues,

> It is precisely because there is in our society no established way of deciding between these claims that moral argument appears to be necessarily interminable. From our rival conclusions we can argue back to our rival premises; but when we do arrive at our premises argument ceases and the invocation of one premise against another becomes a matter or pure assertion and counter-assertion. (8)

In other words, we are left yelling "increased yield" and "endangered water resources" at each other until one side wins its point through sheer volume and tenacity, what MacIntyre characterizes as "the slightly shrill tone of so much moral debate" (8).

To argue in this way can feel as though one's opposition is not behaving rationally. If I insist that studies linking hydrofracking to cancer suggest we should hold off on the process to avoid the risk of premature death to those who live in fracking regions, and my opponent keeps saying "but it will bring good-paying jobs to desperate rural people," my opponent can easily seem to me like an idiot—you can't a hold a well-paying job if you're dead. MacIntyre's point, however, is that each of us *is* thinking rationally about the

situation, but that our rationalities stem from different traditions and contexts that render them "conceptually incommensurable." In *Whose Justice?
Which Rationality?* he writes: "Arguments . . . have come to be understood
. . . not as expressions of rationality, but as weapons" (5). But for MacIntyre,
"since there are a diversity of traditions of enquiry, with histories, there are,
so it will turn out, rationalities rather than rationality" (9).

Parts II and III of this book explores the ways we construct places and our
relationships to them rhetorically; it also helps us see the competing rationalities that underlie these rhetorics. When Massachusetts state representative
Marta M. Walz says, in Chapter 5, that "the Longfellow Bridge is one of
my oldest and neediest constituents," we see underlying Leopoldian values
of "love, respect, and admiration for land" (223, quoted in Kirsch, this volume), but we also glimpse an underlying rationality that implies that based
on these feelings, we should not substantially alter this "iconic" structure.

In Chapter 6, Matthew Ortoleva describes *The Narragansett Dawn*'s
fusion of Christian and Narragansett spiritual traditions to create an ethic
of eco-centrism. In MacIntyre's analysis, this need to blend rhetorical tropes
from two distinct traditions reflects an underlying "epistemological crisis."
In *Whose Justice? Which Rationality?* such a crisis occurs when "conflicts
over rival answers to key questions can no longer be settled rationally"
(362), or, in other words, when the traditional rational structures no longer
work. In Margret Carter's piece "The Soul of the Indian" in *The Narragansett Dawn*, she states, "The Indian had many rational explanations for
his religious attitude, but because he made no separation between his religious and his daily life, one does not deem his explanations always entirely
satisfactory." MacIntyre argues that an epistemological crisis "requires the
invention or discovery of new concepts and the framing of some new type
or types of theory" (362), and we see this in Carter's careful selection of the
term "mysteries." As Ortoleva notes, with its echoes of both native Narragansett spirituality and Catholicism, the term lets Carter fuse Narragansett
rationality with Catholic rationality, the kind of "imaginative conceptual
innovation" (362) MacIntyre asserts as necessary for a tradition to weather
an epistemological crisis. Carter's strategies allow not only for a broadened
rhetorical appeal, but also for the conversion of old rational structures into
a rationality compatible with her present circumstances.

Chapter 9 in this volume provides another example of a blending of
rationalities. In her analysis of the *Atlas of the Patagonian Sea*, Amy
D. Propen concludes that "the *Atlas* draws on rationales for conservation that are rooted in combinations of ethical, utilitarian, ecological,
and aesthetic attitudes." How can we know whether this is an attempt
to combine conceptually incompatible traditions or the emergence of a
transformed tradition from an epistemological crisis? Only time will tell,
but my point, rooted in MacIntyre's analysis, is this: we must come to
see the factions in environmental debates not only as occupying different
rhetorical stances, but different rational traditions, if we are to have any

hope of reducing the "interminable" and "shrill" character of so many environmental debates.

Both Cynthia Haller, in Chapter 7, and Becca Cammack, Linn K. Bekins, and Allison Krug, in Chapter 12, note the prevalence of appeals to *ethos* in ecological controversies. Haller writes, "With logical appeals to sustainability capable of supporting very different policy goals, the importance of appeals to *ethos* increases. The public's identification with a particular side of a controversy may well be determined by which side successfully creates an *ethos* of acting in the interests of sustainability." In other words, when rationality cannot serve to distinguish a preferred position, *ethos* becomes a determining factor. What Haller describes here echoes MacIntyre's description in *After Virtue* as "the culture of emotivism":

> Emotivism is the doctrine that all evaluative judgments and more specifically all moral judgments are *nothing but* expressions of preference, expressions of attitude or feeling, insofar as they are moral or evaluative in character . . . [I]n the realm of fact there are rational criteria by means of which we may secure agreement as to what is true and what is false. But moral judgments, being expressions of attitude or feeling, are neither true nor false; and agreement in moral judgment is not to be secured by any rational method, for there are none. (11–12; emphasis in original)

MacIntyre argues that within the culture of emotivism, "characters" serve as "the masks worn by moral philosophies" (28). MacIntyre means "characters" in the sense of "stock characters" (27); recognizable types that embody "those social roles that provide a culture with its moral definitions" (31). We can see the possible links here with rhetorical appeals to *ethos*; where MacIntyre describes such characters as the Rich Aesthete, the Manager, and the Therapist, we might imagine such stock figures as the crying American Indian (played by an Italian-American actor) of 1970s anti-pollution advertisements, or the "smart diversification" farmer Eileen E. Schell describes in *Rural Literacies*, also embodied in the curriculum Haller describes in Chapter 7.

MacIntyre writes that "emotivism is a theory embodied in *characters* who all share the emotivist view of the distinction between rational and non-rational discourse" (30); furthermore, these characters "are partners as well as antagonists"—they work together to limit "the modes of social life open to us" (35). One might also say they work together to limit the sorts of relations we can have to places. For example, in Chapter 10 of this collection, Samantha Senda-Cook and Danielle Endres show how the "place of one's own" rhetoric adopted by outdoor recreators operates within the same "nature/culture divide and its problematic assumptions" that can "endanger the very areas that are set aside for preservation." Senda-Cook and Endres persuasively argue that the "place of one's own" framework—embodied in the character of the urban outdoor enthusiast who needs to get away from it all to find meaning alone, or with a small group, in unsullied nature—"may be less about developing a sense of place and more about human competition and ownership." One can see how this "character," and that of those who wish to access the natural resources in such areas for economic gain, work as "partners as well as antagonists" to maintain the nature/culture divide.

Because appeals to *ethos*, through the deployment of "characters" in MacIntyre's sense, can be so limiting, what remedy do we have? Must we simply cede the field of *logos*, if the rationalities and values underlying opposing viewpoints are conceptually incommensurable?

The first epigraph to Peter Goggin's introduction to this volume offers one clue. David Orr writes,

> A place has a human history and a geologic past: it is a part of an ecosystem with a variety of microsystems, it is a landscape with a particular flora and fauna. Its inhabitants are part of a social, economic, and political order: they import or export energy, materials, water, and wastes, they are linked by innumerable bonds to other places. A place cannot be understood from the vantage point of a single discipline or specialization. It can be understood only on its terms as a complex mosaic of phenomena and problems. (129)

A place, in other words, is shaped by a multitude of competing rationalities and rhetorics, with complex partnerships and antagonisms among them. For Orr's term "vantage point," we might substitute MacIntyre's use of "standpoint." In "A Partial Response to my Critics," MacIntyre writes that "from the standpoint of each particular tradition . . . relativism fails, just as do the contentions of its rivals" (294). In other words, from a single vantage point, there are not "rationalities," but rather one superior rationality—that of one's own tradition. One's "rivals'" viewpoints are inferior. But as Orr suggests, places—in the sense of Yi Fu-Tuan's definition as spaces invested with meaning (8)—are subject to a multiplicity of traditions to derive that meaning. How can we understand places without inhabiting multiple standpoints? And how can we inhabit multiple standpoints without falling into relativism, as MacIntyre devoutly wishes to avoid, and losing all sense of our particular convictions that shape our personal relationships to places?

It's a tall order. In *Whose Justice? Which Rationality?* MacIntyre describes a two-stage process to resolve "genuine controversy" between "rival intellectual traditions":

> The first is that in which each characterizes the contentions of its rival in its own terms, making explicit the grounds for rejecting what is incompatible with its own central theses, although sometimes allowing

that from its own point of view and in light of its own standards of judgment its rival has something to teach it on marginal and subordinate questions. A second stage is reached if and when the protagonists of each tradition . . . ask whether the alternative and rival tradition may not be able to provide resources to characterize and to explain the failings and defects of their own tradition more adequately than they, using the resources of that tradition, have been able to do. (166–167)

This prescription—to criticize one's rival with the rival's own terms, within the rival's own system, and to probe one's own tradition for its inadequacies—sounds much like the call for deliberative rhetoric Peter Goggin makes in the introduction to this collection:

Deliberative rhetoric . . . does not presume to seek consensus per the classical understanding of deliberation, but rather offers such a reframing of place and environmental discourse by encouraging stakeholders to articulate opinions and beliefs as a means to critically examine them (and those of others) to foster effective argumentation, change, and synchronicity.

MacIntyre's perspective means that we cannot take lightly the task of critically examining not only opinions and beliefs but also foundational assumptions and rationalities—our own and others'. It is a process, writes MacIntyre, that "requires a rare gift of empathy" (167) to truly get inside the intellectual tradition of another to understand the advantages and disadvantages of their system. As the essays in this collection show, it is a difficult task worth pursuing, as the ways we make places meaningful to us profoundly affect the futures of those places and the people who inhabit them.

WORKS CITED

MacIntyre, Alasdair. *After Virtue: A Study in Moral Theory*. Second edition. Notre Dame, IN: U of Notre Dame P, 1984. Print.
———. "A Partial Response to my Critics." *After MacIntyre: Critical Perspectives on the Work of Alasdair MacIntyre*. Ed. John Horton and Susan Mendus. Cambridge: Polity, 1994. 283–304. Print.
———. *Whose Justice? Which Rationality? A Study in Moral Theory*. Notre Dame, IN: U of Notre Dame P, 1988. Print.
Orr, David. *Ecological Literacy: Education and the Transition to a Postmodern World*. Albany: State U of New York P, 1992. Print.
Schell, Eileen. "The Rhetorics of the Farm Crisis." *Rural Literacies*. Kim Donehower, Charlotte Hogg, and Eileen E. Schell. Carbondale: Southern Illinois UP, 2007. Print.
Tuan, Yi-Fu. *Space and Place: The Perspective of Experience*. Minneapolis: U of Minnesota P, 1997. Print.

Contributors

Linn K. Bekins is Associate Professor of Professional Communication and faculty at the Center for Bio/Pharmaceutical and Biodevice Development at San Diego State University. She earned her doctorate in Educational Studies and Rhetoric and Composition at the University of Utah with specialties in collaboration and scientific writing. She has consulted extensively or provided training with pharmaceutical, applied science, and healthcare companies interested in improving communication practices between scientists and the general public. At SDSU, her primary research interests focus on the intersection of written communication, learning, science, and organizations, and she has published in each of these areas.

Elizabeth A. Brandt is a socio-cultural and linguistic anthropologist. She is a Professor of Anthropology in the School of Human Evolution and Social Change at Arizona State University. She is also affiliated faculty with the PhD program in Applied Linguistics, Women and Gender Studies, and American Indian Studies. Her research interests are in senses of place, the use of cultural and natural resources and their protection, sustainability, gender, and language and cultural preservation. She has served as an expert witness in land claims and in issues over contested places. Her most recent publication is a book chapter coauthored with Steven Semken, "Implications of Sense of Place and Place-Based Education for Ecological Integrity and Cultural Sustainability in Diverse Places."

Becca Cammack has worked in the environmental and energy industries since graduating with her bachelor's of science degree in Geology from Northern Arizona University in 2000. After spending over five years working in various roles and disciplines in the environmental consulting industry, she joined the Southern California energy industry as an environmental professional specializing in water quality. She most recently worked as a Senior Environmental Coordinator for Pacific Gas & Electric, Co. in Northern California, where she supported project teams in their efforts to understand and comply with environmental permits and regulations. Becca is certified in Technical and Scientific Writing and is completing her master's of arts degree in Rhetoric and Writing Studies at San Diego State University.

Rick Carpenter is Associate Professor of English at Valdosta State University in Valdosta, Georgia, where he teaches undergraduate and graduate courses in writing, rhetorical theory, new media, and composition pedagogy. He has published in *M/C Journal: A Journal of Media and Culture*, *Computers and Composition*, and *Disability Studies Quarterly*, and has chapters in the forthcoming collections *New Media Literacies and Participatory Popular Culture Across Borders* and *Disrupting Pedagogies and Teaching the Knowledge Society*. His research interests include genre theory, new media studies, and identity construction, particularly in relation to place.

Brian Cope's writing, teaching, and scholarship is fueled by his active involvement in numerous campaigns to protect western Pennsylvania forestlands and watersheds. As a doctoral candidate in the Composition and TESOL program at Indiana University of Pennsylvania, Brian Cope's dissertation, "The Ecologic Episteme: A Pathway to a Literacy of Sustainability," is a historically contextualized study on ecologic thinking in composition and writing studies. His essay on environmental perception in western Pennsylvania was published in *Literature, Writing, and the Natural World* (2010). He has presented his eco-English research at national and regional conferences, including CCCC, ASLE, and WSRL.

Kim Donehower is Associate Professor of English at the University of North Dakota, where she researches the relationship between literacy and the survival of rural communities. With Charlotte Hogg and Eileen E. Schell, Kim coauthored *Rural Literacies* (SIUP) and coedited *Reclaiming the Rural: Essays on Literacy, Rhetoric, and Pedagogy* (SIUP). Her essays have appeared in *Women and Literacy: Local and Global Inquiries for a New Century*, *Rethinking Rural Literacies: Literacy/Rurality/Education*, and *Literacy, Economy, and Power: New Directions in Literacy Research*.

Danielle Endres is an Assistant Professor of Communication and faculty in the Environmental Humanities master's program at the University of Utah. Her research focuses on the rhetoric of environmental controversies and social movements including nuclear waste siting decisions, climate change activism, and energy policy. Endres' secondary interest is in the rhetoric of American Indian activism. Endres is the coeditor of *Social Movement to Address Climate Change: Local Steps for Global Action* and has published in *Quarterly Journal of Speech*, *Communication and*

Critical Cultural Studies, *Western Journal of Communication*, *Environmental Communication*, and *Local Environment*.

Peter N. Goggin is Associate Professor of English (Rhetoric) at Arizona State University where he studies and teaches theories of literacy, environmental rhetoric, and sustainability. He is the editor of *Rhetorics, Literacies, and Narratives of Sustainability* (2009) and author of *Professing Literacy in Composition Studies* (2008). His articles on literacies of sustainability, environmental rhetoric, and environmental discourse, rhetoric, and writing include publication in *Composition Studies, Community Literacy Journal*, and *Computers and Composition*. He is a Senior Scholar with ASU's Global Institute of Sustainability, and his current research includes the study of rhetorics and discourses of sustainability and globalization in oceanic islands. In addition to Arizona he has taught graduate and undergraduate courses and seminars in Romania, China, Bermuda, Boston, Pittsburgh, and Austria. He is founder and codirector of the annual Western States Rhetoric and Literacy conference, which features themes on sustainability, culture, transnationality, and place.

James Guignard is Associate Professor of English and Director of Composition at Mansfield University. He teaches composition, professional writing, environmental literature, and mountain biking. He lives with his family in the midst of the Marcellus Shale development where he has watched machines move with alacrity into the garden of Tioga County, Pennsylvania. Even so, he still believes Tioga County is a wonderful place to teach, raise kids, and ride bikes. For now.

Cynthia R. Haller is Associate Professor and Deputy Chair of English at York College, City University of New York. Her research on scientific and technical communication has appeared in journals such as *Written Communication, Journal of Engineering Education*, and *Technical Communication Quarterly*, where she published an article on reading and writing in the North Carolina Girls' Canning Clubs. More recent research, published in *Writing Program Administration* and in a forthcoming edited volume on reading, has focused on students' development of academic literacies. She is currently investigating rhetorics of agriculture, agriculture in literature, and environmental rhetoric.

Gesa E. Kirsch is Professor of English at Bentley University, Waltham, MA. Her research and teaching interests include feminism and composition, ethics, qualitative research methodology, archival research, women's rhetorical education, and environmental rhetoric. In 2011, she participated in a NEH Summer Institute on Sustainability and the Humanities. Her books include *Feminist Rhetorical Practices: New Directions for Rhetoric, Literacy, and Composition Studies* (2012), coauthored with

Jacqueline Jones Royster; *Ethical Dilemmas in Feminist Research* and *Women Writing the Academy*; and five edited collections, most recently, *Beyond the Archives: Research as a Lived Process* with Liz Rohan.

Allison Krug is an epidemiologist and biomedical communications consultant. She earned her master's in public health with concentrations in Epidemiology and Health Economics from the University at Albany (SUNY) School of Public Health. She also holds an advanced certificate in Scientific and Technical Writing from San Diego State University. She has served as coauthor and editor for manuscripts published in public health and social science journals related to pharmacoeconomics, infectious disease epidemiology, health care quality, and environmental policy.

Jaqueline McLeod Rogers is an Associate Professor in the Department of Rhetoric, Writing and Communication at the University of Winnipeg. She has published on place in a recent special issue of *Writing on the Edge* ("Writing Winnipeg, Manitoba, Canada: Dwelling and Crossing") and is continuing to research the connections between rhetoric and geography. She recently co-convened a conference on Marshall McLuhan and is currently preparing a paper comparing how McLuhan and Jane Jacobs viewed city and culture. She has published on Margaret Mead and family rhetorics ("Moms and Teen Daughters: Make Room for the Internet"). Linking her research interests is the question of how we form and strengthen community.

Matthew Ortoleva is currently Assistant Professor of English and Writing Center Director at Worcester State University. He studied rhetoric and writing at the University of Rhode Island and earned his PhD in English in 2010. His dissertation, titled "Rhetorics of Place and Ecological Relationships: The Rhetorical Construction of Narragansett Bay," is an ethnographic study of the discourses that construct, challenge, and change ecological relationships with the Narragansett Bay Watershed. This essay grows out of his work with Paulla Dove Jennings, a Narragansett elder, on spiritual relationships to Narragansett Bay. His work has also appeared in the *Community Literacy Journal*.

Rebecca Powell is completing her doctoral studies at New Mexico State University in literacy, composition pedagogy, and culture criticism. Before graduate school, she covered growth and land development issues in the West for online and print publications. Currently, she is researching the connections between writing and spatial literacy. Her other publications focus on teacher inquiry and narrative, online pedagogy, and mommy blogging. When she's not teaching and writing, she enjoys hiking and bicycling the Organ Mountains.

Amy D. Propen is a Lecturer of Rhetoric and Composition in the Writing Program at the University of California Santa Barbara. Her research on visual rhetoric, critical cartographies, and rhetoric as advocacy has appeared in *Technical Communication Quarterly, Journal of Business and Technical Communication, Written Communication, Law, Culture and the Humanities, ACME: An International E-Journal of Critical Geographies*, and the edited collection *Rethinking Maps: New Frontiers in Cartographic Theory*. She is coauthor, with Mary Lay Schuster, of *Victim Advocacy in the Courtroom: Persuasive Practices in Domestic Violence and Child Protection Cases*, and author of *Locating Visual-Material Rhetorics: The Map, the Mill, and the GPS*.

Sally Said is Professor of Modern Languages at University of the Incarnate Word in San Antonio, Texas, where she teaches Spanish and linguistics. Her research interests include eco-rhetoric, writing pedagogy, gender issues, Spanish dialectology, and Navajo culture. She has coedited proceedings for two international women's conferences, and has published a collection of Sudanese folktales as well as contributions to anthologies, including a chapter examining the rhetoric of the Headwaters Coalition, which oversees a fifty-three-acre nature sanctuary on the San Antonio River adjacent to the UIW campus. She spends half of each summer at Diné (Navajo) College in Tsaile, Arizona.

Samantha Senda-Cook is an Assistant Professor of Communication Studies at Creighton University. Within the field of communication, Samantha studies rhetoric and environmental communication. She's particularly interested in how authenticity functions as a rhetorical strategy to shape humans' relationships with the natural world. Her current research investigates the connections between masculinity and perceptions of survival in outdoor recreation, intersections of "real nature" and homelessness in urban parks, and ethnographic approaches to rhetorical analysis. Senda-Cook has published in *Quarterly Journal of Speech* and *Western Journal of Communication*.

Michael Springer is an adjunct instructor of English and communications at Bryan University. He maintains the blog *Humanities for the Environment*, part of an interdisciplinary initiative focused on the humanities' role in sustainability discourse. Michael holds an MA in rhetoric and composition from Arizona State University, and his research interests include video game eco-criticism, social networks in education, and the role of the humanities in science communication.

Deborah L. Williams is a socio-cultural anthropologist and PhD candidate at Arizona State University. Her research interests include place theory; the intersection of place, identity, and heritage; place-based discourse and cultural preservation; environmental justice; and global health. Her current research examines culture-specific meanings and how these are discursively recast in the multicultural arena surrounding contested places. In addition to her current work on the rhetoric of contested place, she has published on ethnographic methods in assessing place-based geosciences education. Her most recent publication was coauthored with Dr. Steven Semken and titled "Ethnographic Methods in analysis of Place-Based Geoscience Curriculum and Pedagogy."

Index

A

active society, 190–192, 196
adaptive capacity, 56
AFBFA (American Farm Bureau Foundation for Agriculture), 100, 101, 102
agribusiness, 37, 98
agricultural industry, 98–99, 100–101, 104
agricultural literacy, 97–100, 105, 106
agriculture: as sustainable, 37, 104; attitudes towards, 97–100, 103, 106; literacies of, 98–100, 105–106, 108; policies, 97, 99
AITC (Agriculture in the Classroom), 99, 100, 105
Anthropocene Epoch, 2
Appalachia, 28, 29, 31, 36
authenticity: in experiencing nature, 144, 145, 158; in gaming environments, 114

B

biomass, 10, 199, 200–213
biotic community, 7, 70, 78, 88
Boston, 9, 69–81

C

Cape Wind project, 201
carbon dioxide 30, 201
Caribbean, 4, 111
cartography, 8, 127, 128
CCNGD (Citizens Concerned about Natural Gas Drilling), 23–25
Cebolleta Land Grant, 58, 62–63
CGP (Construction General Storm Water Permit), 177–184
Charles River Esplanade 71, 77, 78, 81
Charles River, 70–71, 74–75, 77

Chesapeake Energy, 15, 18, 25
chora, 238
Cibola National Forest, 60
civic participation, 10, 71, 160, 186–187, 196–197
climate change, 2, 3
coal, 28–30, 32, 37–39, 115, 201
colonialism, 2
community visioning project, 78
complexity theory, 3–4, 29, 34, 55–56
conceptual incommensurability, 218–220, 222
consensus (and dissensus), 6, 59–61, 181, 187, 189, 192, 196, 203, 223
conservation, 4–9, 29, 42, 45–46, 56–57, 103, 107, 115–116, 120, 127–129, 134–138, 184, 190–191, 201, 217, 220
counterpublics, 22–26, 84–86, 94
CPRC (Cultural Property Review Commission), 54, 58, 61–64
cultural geography, 8, 159
culture of emotivism, 221

D

Dawes Act (1887), 85
degradation, of places and environments, 10, 50–51, 150–151, 174, 176, 180, 207
deliberative rhetoric, 6, 10, 223
Des Moines, 3
Devil's Canyon, 46
diachronic structure, of ideographs, 103–104, 107
digital: environment, 111; gaming,111; representation, 9, 114, 121; space, 118; text, 127–128
discourse: competing and divisive, 5, 10, 200; contested and counter,

84, 128; of culture and identity, 84, 86, 137–138, 164, 228; ecocentric, 9, 86–87, 94; environmental, 6, 71, 102–103, 173–174, 186–187, 196, 199, 207, 217, 223, 227; hegemonic, 2, 6, 84, 206; industrial, 210–212; of instrumental rationality, 107; of outdoors and recreation, 143, 145–148, 150–152; of politics and policy making, 8, 47, 71, 107; as provocative lens, 10, 221; public, 16, 23, 57, 69, 71, 74, 80, 196; regulatory, 8, 10, 25, 200–204; of science and technology, 138, 201, 202; shared, 23, 209; spiritual, 9, 91–92; of sustainability, 103, 107
displacing rhetoric, 16
Doña Ana County, 10, 186–188, 191–192, 197
Dubai, 3

E
ecocentricism, 5, 86–89
ecocomposition, 117
ecological: community, 87–88, 90–91; consciousness, 103, 129; damage, 218; debates, 217–218; diversity, 29, 112, 113; and epistemology, 5; ethic, 84, 199, 220; identity, 9, 86, 92, 138; images, 35; literacy, 3, 86, 118, 207; perspective, 137, 207, 211–212; research, 7; resilience, 55; as rhetorical, 28; systems, 5–6, 9, 55–56, 70, 207; sustainability, 6, 150, 180
ecology: definition of, 6–7; of the classroom and education, 97, 105–106; and epistemology, 5; of the farm, 105; as metaphor, 7; of situations, 195; of urban settings, 155; of video games, 112; of writing, 207
econ, 30, 35–36
ecospeak, 5, 102–103, 107, 175, 200
educational games, 100, 113–114, 121
Effra Road 161
elephant seal, 134–138
emplacing rhetoric, 16
Environmental Assessment, 81

environmental rhetoric, 4–6, 71–72, 80, 106, 127, 134–138, 199, 201
environmental: impact, 4, 17, 21, 55, 114, 116–117, 121, 204; racism, 204–205, 208; regulations, 8, 10, 22, 63, 173–179, 181, 184, 211, 219
environmentalism, 6, 46, 98, 145, 173–175, 199
EPA (Environmental Protection Agency), 23, 25, 177, 183
ethos, 33, 85, 97, 106, 107, 173, 193, 201, 221, 222
exigency, 30, 34–39, 188–189
extraction process, hydrofracturing, 16, 31

F
false exigency, 29, 31, 33, 35, 38, 219
family farms, 35, 37, 98
FRAC Act, 21
fracking (hydrofracturing), 8, 16, 20–21, 24–25, 28–39

G
Games for Change Festival, 112–113
gas industry framing metaphors: as expert, 20, 21; as infant, 20, 21
Gasland, 30, 32
geographic rhetoric, 10, 200, 202–212
geopolitics, 2
GIS (Geographical Information System) 129, 132–135
global: communication networks and systems, 3, 4,, 5, 6, 8 15, 43, 134; culture and perspective, 4–6, 161, 166–168; economics and trade, 2, 3, 4, 48; environmental health, 115–116, 121; solutions, 3, 183
globalization, 2–5, 112, 121, 167
glocalization 5–6, 8
God-terms, 103, 107
Grants, New Mexico 8, 54–55, 59

H
habitus, 164
heterotopias, 127–129, 138
hikers and hiking, 10, 145, 151, 179, 180

I
ideographs, rhetorical dimensions of, 97, 103–104, 106–107

immersive game play, 112–114, 117–118
instrumental rationality, 107

J
Johnstown, 18
just war, notion of a, 217–218
justice: environmental, 95, 204; social, 5, 8, 62, 160, 218, 220, 222

L
land ethic (Aldo Leopold), 69–82
landscape: aesthetic and recreational value of, 51–52; culture and identity, 5, 7, 42, 44–45, 48–51, 57, 187, 202, 212, 222; digital and futuristic types of, 120; in images and imagery, 4, 15, 25; linguistic, 156, 165–166; urban, 10, 112, 155, 158, 165
land-use planning 187
language: and animals, 137; in culture 80, 145, 194, 195; of industry, 16–19, 20, 210; as persuasive, 102, 107, 174, 175, 182; as place-based practice, 70, 97, 156, 164–166, 199; for protest, 23; for regulatory purposes, 177, 178, 183
Las Cruces, 187, 188, 189, 191
Latinos/Latinas, 42–43, 45, 47–51
localization and localism, 5, 16, 37, 48, 166, 199, 207, 209, 212
logos, 173, 222
London, 3, 161, 166
Longfellow Bridge, 61–82
Los Angeles, 3

M
Magma Mine Company, 42, 45, 49
Marcellus Shale Coalition, 17–18
Marcellus Shale, 8, 15–26, 28–39, 217
marine ecosystems, 129–130
Mars rover (Curiosity) 1–2
Mars, 1–2
MassDOT (Massachusetts Department of Transportation), 69, 71, 73–81
material rhetoric, 9, 118, 127–129, 134, 137–139
material: effect, 17, 157, 212; literacy, 100, 105, 187, 194, 207; presence in nature, 2, 144, 151, 206–208; policy, 102; space, 1,

127, 155, 159, 162, 164, 203, 205
metanarrative, 103
minero identity, 47, 50–52
mobility 62–163, 167–168
moral reasoning, 10, 103, 217, 219, 221

N
Narragansett Bay, 9, 84–86, 89–90
narrative, 6, 37, 44, 48, 51, 87, 89, 118, 121, 138, 151, 161, 168, 175, 183, 186
national security, 18
nationalist rhetoric, 209
Native American cosmologies, 57
Native American environmentalism, 84
Native American people: Acoma Pueblo, 54, 55, 56, 58, 61; Apache, 42–43, 46, 49, 56, 61; Hopi, 54–57, 61; Laguna Pueblo, 54–55, 57–58, 61; Narragansett, 9, 84–95, 220; Navajo, 54–55, 57–58, 61; Yavapai, 42–43, 49; Zuni, 54–56, 61
natural gas: estimated volume of Marcellus Shale, 15, 31; development, 8, 16, 18–19, 21, 23–24; drilling and extraction, 18, 23; industry, 16, 20, 25
natural world, the, 84, 88–94, 129, 136, 143–147, 150–151, 163
nature/culture divide, 10, 143–147, 150–151, 221–222
neighborhoods, 4, 74, 80–81, 156, 188, 191, 194, 196, 208–209
neighbors, 70, 175, 184

O
Oak Flat, 42, 46, 49
Oakland, 1

P
paradigm of sustainability, 102–104, 107, 219
pathos, 30, 36, 37, 173
pedagogy, 9, 105–106, 111, 113, 199
Pennsylvania regulation HB 1950 on hydrofracking zoning, 33
philosophy: ecological (internal relatedness), 86; western history of, 2; video gaming, 111
PIOGA (Pennsylvania Independent Oil and Gas Association), 20

place: ecological definition of, 7; economics, 28, 31, 190, 191; as escape/refuge, 143–152; as experience, 1, 156–169; as intrinsic value, 3, 75, 143, 144–154; as memory, 1, 2, 47, 56–57; as myth, 2; and publics, 23, 24, 70; and relationship with space, 7, 8, 17, 20, 199, 202–205, 207, 208–210, 213; as representation of identity, 1, 3, 36, 37, 42–44, 47–49, 51, 97–100, 155, 186–197, 207, 212, 213; as rhetorical construct, 1–10, 29–31, 38, 45–52, 70, 75, 78, 80, 111, 118, 201–202, 220–223; as spiritual connection, 84–95; and sustainable futures, 5, 6, 97, 100–105; as symbolic meaning, 97; as transformative, 2, 38, 44; as virtual experience, 11–123
place-based rhetoric, 5–10, 46, 111, 186, 208
place-based: arguments, 45, 190, 212; learning 9
politics of space, 200–204
pollution: air, 8; ground, 3, 17, 50, 115; sources of, 174, 177, 180, 181; water, 25, 37, 58, 174, 177–178, 181–183
praxis, 2, 175
presence: as liminal/marginal, 168; and material rhetoric, 2, 118; in nature, 145–146, 150–151
preservation: cultural, 43, 47, 191, 217; environmental, 4, 6, 143, 145, 221; historic, 54, 58–63; and material rhetoric, 127
President Barack Obama, 25
President Franklin D. Roosevelt, New Deal, 85
President Zachary Taylor, 55
procedural rhetoric, 114
Promethean epistemology, 3
property rights and ownership, 46, 50, 54, 57, 59, 62, 85, 190, 200, 205, 209
publics, 22, 84, 189, 190, 200, 202, 212

Q
Queen Creek, 46, 49

R
Range Resources, 15, 28, 31, 33, 34
RCC (Resolution Copper Company), 42, 46, 47, 49, 51
regional rhetoric, 4, 6
regional: blogs, 24; planning, 10, 187–189; nesting sites, 131; networks, 70; resistance, 159
religion, 2, 34, 49, 57, 63, 89, 92–94, 220
renewable energy, 38–39, 116, 201
resilience: ecological and social, 54–65; and sustainability, 4, 75
reverence (and irreverence), 28–39
rhetoric: of environmentalism, 71; of presence 4, 5, 182, 183
rural identity and culture, 22, 49, 97–98, 159, 191, 218–219
rural rhetoric 4, 8
rurality 4

S
San Diego, 10, 177
satellite tracking (of wildlife), 2, 9, 127, 131–140
sense of place, 1, 2, 5, 7–10, 17, 24, 42–52, 76, 88, 143–145, 148–151, 207, 208, 222
severance tax on resource extraction, 20–21
Sierra Club, 46, 58, 63
South Georgia Island, 130–132
spheres: global, 2; local, 2; public, 5, 84, 118, 138, 167; social, 118, 138, 167
spirituality, 8, 9, 29, 55, 84, 86, 89–94, 173, 218, 220
stewardship, 101–107, 219
storm water runoff, 77, 174–175, 177–178, 180–183
Superior, 8, 42–45, 47–52, 217
sustainababble, 175
sustainability: and community health, 76–78, 115; ecological, 6, 100–105; environmental, 8–10, 71, 112, 180; and environmental regulation, 173, 176, 180, 181; and environmental rhetoric, 4, 75, 107, 121–122, 174–175, 178, 202, 219–221; paradigm, 102–104, 107, 219; and pedagogies, 111, 113–114, 122, 181; socio-cultural, 10, 72, 174, 178, 184; terminology, 4, 10, 107, 175, 219

SWRCB (California State Water
Resources Control Board), 177,
182

T
TCP (Traditional Cultural Property),
54, 56–59, 61–64
thirdspace, 159, 160
Tioga County, 16, 23
to propen, 195
topoi, 38, 118
transportation modes, 2, 70, 79, 115,
163, 164, 191, 202

U
United States Forest Service, 42, 46,
55, 58, 61, 64
uranium: deposits, 9, 55, 61–63, 217;
mining companies, 54, 55, 57,
63; pollution 55–59

urban: advertisement design, 189;
environment and perspective, 4,
8–10, 70, 74, 99, 155–157, 222
USDA (United States Department of
Agriculture), 99

V
Valdosta, 10, 199–213

W
wandering (about), 122, 163
Wandering Albatross, 129–134
WCS (Wildlife Conservation Society),
127
web of language, 16, 25
Westphalian war system, 3
wilderness: concept of, 19, 69, 143,
152; designated areas of, 112
Winnipeg, 165
wisdom, 56, 89, 92